CW01187415

Gone to ROTTNEST

ROTTNEST ISLAND

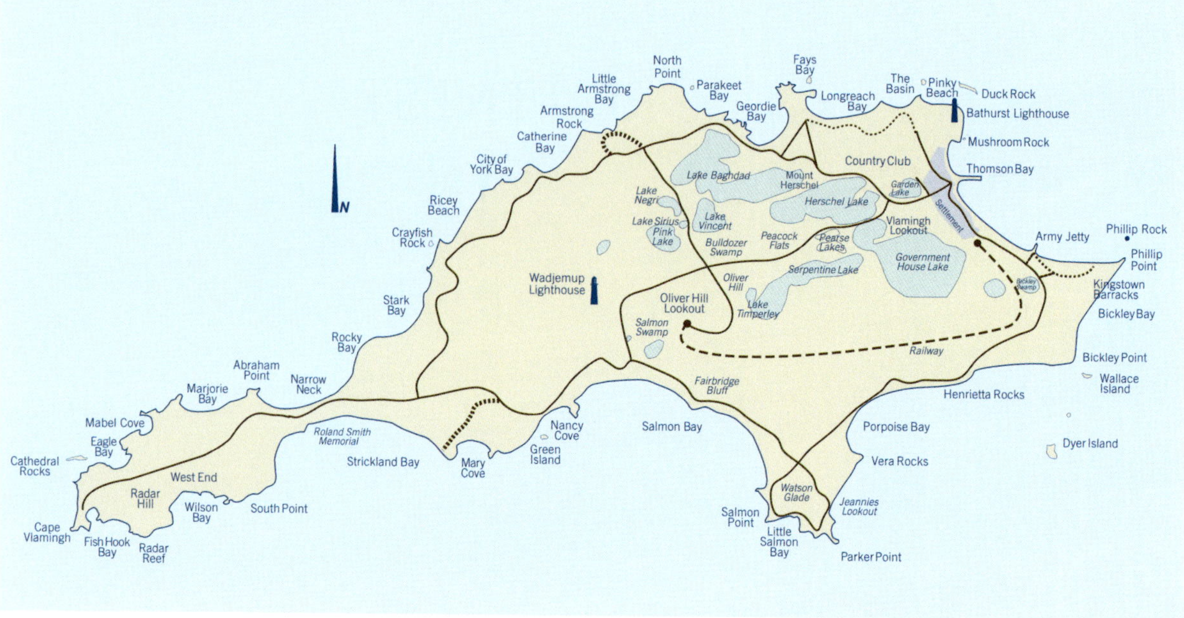

Gone to ROTTNEST

TREA WILTSHIRE

University of Western Australia Press

First published in 2004 by
University of Western Australia Press
Crawley, Western Australia 6009
www.uwapress.uwa.edu.au

National Library of Australia
Cataloguing-in-Publication entry:

Wiltshire, Trea.
Gone to Rottnest.

ISBN 1 920694 28 5 (pbk.).

1. Rottnest Island (W.A.) - History. I. Title.

994.12

Front cover photo by Micheal Chan, 2004; title painting by Nellie Crawford; back cover photo by Pat Barblett

Consultant editor: Ross Haig, Perth
Designed and typeset by John Douglass, Brown Cow Design, Perth
Typeset in 10pt Optima
Printed by Craft Print International, Singapore

CONTENTS

1. AN ISLAND, SOMEWHERE...

FOR most of us, for most of the year, it is merely a smudge on the horizon, lying on the very edge of our collective consciousness. Unsubstantial. Sometimes barely discernible. But come holiday time, it begins to define itself against the blue clarity of a wide sky and glittering ocean. And it beckons...

On the quayside at Fremantle, gulls wheel, fishing boats head for the ocean, wind stiffens the sails of yachts. Bicycles are stacked on waiting ferries poised to cover the 17 km trip in half an hour, baggage crates are lowered onto the decks, and families, groups of teenagers, English backpackers and Japanese day trippers disappear into the sleek interiors.

Only when the ferry is close enough for you to hear the cries of gulls, and smell the salty tang of beach and coastal heath, do you catch your breath—yet again—at the colour and clarity of the ocean. Forget goggles and snorkel, you can see the intricate patterns of the underwater world even from the side of the boat!

As you disembark, you savour what lies ahead—for the particular pleasures of this island keep drawing you back. There will be long walks and bike rides to deserted white-sand coves where you will swim, snorkel and watch birds; there will be fishing—and fish and chips on the jetty; there will be afternoon naps while a busy Welcome Swallow builds its nest in the eaves of your cottage and the sounds of children on the beach are carried on the wings of a sea-breeze.

If it's summer, there may be a late afternoon stroll to a favourite sunset spot, or a sundowner on the veranda, with the bay stretched out before you. If it's winter, there could be a fire, and perhaps a fiercely contested game of Happy Families. Whatever the season, the night air will bring the voices of children

Popular Rottnest may be Western Australians' favourite holiday island, but it is still possible to enjoy solitary walks on deserted beaches. Photo by Pat Barblett.

A Fairy Tern flying solo against the merged blue of sea and sky captures the sense of freedom and escape from routine that is a vital element of any trip to Rottnest. Photo by Pat Barblett.

conversing with nocturnal visitors: the quokkas that boldly hop into the yard foraging for scraps.

On arrival, you will shed your mainland angst and preoccupations and will instantly become aware of the smiles that surround you. Every night you will fall asleep to the sound of the sea and savour the particular pleasure of being on an island.

Welcome to Rottnest!

2. 'THE PLACE ON THE OTHER SIDE OF THE WATER'

WESTERN Australia is blessed with 12,000 kilometres of coastline—from the mangrove-fringed turquoise waters in the northern tropics, to the awesome cliffs that confront the steel blue of the Southern Ocean, rolling up from Antarctica.

However, Australia's largest State is singularly lacking in offshore islands adjacent to its most populous area, the capital city of Perth and the city port of Fremantle.

Rottnest Island in the State's south-west thus enjoys special status, not only because it is readily accessible, but because it has been designated a holiday island.

One of a chain of three small islands that barely break the majestic sweep of the Indian Ocean, it is just visible from Fremantle, at the mouth of the Swan River that snakes through riverside suburbs to Perth. It is the most westerly, and the largest, of this offshore chain.

Physically, Rottnest is a filigree of limestone headlands, white-sand bays, intricately textured exposed reef platforms and salt lakes that mirror the State's enviable blue skies. It is part of a limestone ridge running parallel to the coastline that was drowned when sea levels rose 7,000 years ago.

Ancient artefacts found on the island attest to the fact that *Wadjemup*—'the place on the other side of the water' as it became known to Aborigines—was once part of the mainland territory of the Nyungar people. Recent archaeological finds of prehistoric stone tools suggest the island could be Australia's oldest human cultural site.

The rising sea water that isolated the island also shaped its destiny—as the first landfall for adventurous seafarers; as a prison island; as a forward

defence in wartime; and finally, as a holiday destination, far removed from the workaday world.

Lying at latitude 32 degrees, Rottnest is a fascinating admixture of the tropical and temperate in terms of flora and fauna. It is washed by the warm waters of the south-flowing Leeuwin Current which has a marked effect on Perth's climate. The tropical cloudbands generated off the north-west coast interact with storm systems coming up from the south-west, providing greater rainfall than coastal locations at a similar latitude enjoy.

The Leeuwin Current (driven by the build-up of tropical water flowing from the Pacific via the Indonesian archipelago) carries a swath of warm water 50 kilometres wide and 200 metres deep, down the coast. During autumn and winter, this significantly warms the waters around Rottnest.

Apart from rewarding hardy winter bathers with temperatures 4 degrees warmer than on the mainland, the Leeuwin sweeps the eggs and larvae of tropical fish from spawning grounds in the north. The conditions the current creates contribute to the attractions of Western Australia: its benign winters, excellent wines and peerless lobsters.

In the early days of the Swan River Colony, attempts were made to turn the island into an extension of the mainland settlement. British settlers shipped over their livestock, grain and furniture, drew up plans for a town called Kingstown and dreamed of transforming the picturesque island into farming estates not unlike those left behind in England.

Later, Rottnest's isolation made it appealing as a prison that promised no escape for those incarcerated—but a delightful escape for colonial governors

In the late afternoon the coastal heath and sheltered coves take on a burnished tone—persuading visitors to return to the emptying beaches to toast the elements of Rottnest with a sundowner. Photo by Micheal Chan.

enjoying its vice-regal retreat, snipe shooting and respite from the summer heat.

In the twentieth century's first decade, attempts were made by parliamentarians to open the island to private ownership, to landscape it, and slice it up to create 400 desirable half-acre blocks. However, by that time enough West Australians had swum in Rottnest's gin-clear waters, caught herring off its rocky outcrops and trekked through its coastal heath, to know that this island was a gem that they had to protect and preserve.

The groundswell of public opinion against private ownership was backed by the governor of the day who declared the island a public park, a place of recreation. Its natural beauty, it was decreed, should remain, as far as possible, undisturbed.

3. FIRST IMPRESSIONS

ALTHOUGH prehistoric Aboriginal artefacts have been found on Rottnest, it is doubtful that anyone lived on the island once it had been isolated by the ocean.

However, in the seventeenth century the lure of eastern spices ensured that the island would soon feature on mariners' charts. Amsterdam's warehouses and the promise of profits from the East Indies spurred Dutch merchants and mariners to sail to the 'spice islands'.

The Dutch were among the first to sight Australia's west coast as they pursued a new route across the Indian Ocean that would cut the twelve-month voyage between Amsterdam and their trading outpost of Batavia (Jakarta).

As early as 1616 Dirk Hartog landed on the island that now bears his name, off the north-west coast of Western Australia. The charts of successive Dutch vessels slowly traced the coastal contours of the land variously called Terra Australis, the Southland and (optimistically) New Holland. Its rugged western coastline, sometimes rearing into spectacular cliffs, would claim many a Dutch vessel.

Rottnest first appeared, unnamed, on the charts of early adventurers in 1627. Over thirty years later, a search party was sent to discover the fate of the Dutch East India Company's *Vergulde Draeck* (Gilt Dragon). The vessel had disappeared with valuable cargo en route to Batavia and was thought to have been wrecked off the coast of Western Australia. When search vessels surveyed the coastline, Rottnest was sighted and the captain of the *Waeckende Boey* anchored half a mile off the island. A boat was despatched, and the ship's log later recorded that the island was well wooded, but surrounded by dangerous reefs.

As the Dutch vessel continued to chart the coast, it encountered fierce storms that appeared to claim a boat and its fourteen-man crew sent to the mainland to look for signs of survivors of the *Vergulde Draeck*. When the weather worsened, the master of the *Waeckende Boey*, fearing the worst, abandoned his search for the lost boat. He sailed for Batavia, unaware that his men had survived and would eventually follow him north, though only four would survive the terrible journey.

The next recorded landfall on Rottnest was made in 1696 by the brilliant cartographer, Willem de Vlamingh, also searching for a lost Dutch East India Company vessel while surveying the coast. Vlamingh's expedition was important to Rottnest in that the island was subsequently named on the charts.

Vlamingh's voyage in the *Geelvinck* (Yellow Finch) had been hard, for he came to Australia's west coast through storm-swept waters, via the southern islands of Tristan da Cunha, St Paul and New Amsterdam. As landfalls, these islands were isolated and inhospitable. Tranquil Rottnest, etched in the blue of the Indian Ocean, must have seemed like a mirage to the weary sailors. Vlamingh's first fanciful description of it as 'the isle of mists' perhaps evokes the summer haze that can dissolve the horizon, making the island appear illusory, suspended between earth and sky.

Vlamingh's vessel anchored and members of the crew were sent to explore, search for signs of shipwrecks, and spend a night on the island. They killed some 'wood rats', cut aromatic branches from trees and returned with stories of the island's benign nature. Some were convinced they had heard a nightingale in the carolling of morning magpies.

Vlamingh's log records:

> On December 31 after hearing so favourable a report, my people prayed me to go with them to see the place and make a more complete discovery. To content my curiosity I went and landed without difficulty...We perceived a very agreeable belt of trees, very thick and about half a league in extent. The trees were placed as if planted in lines by the industry of man to form a park. We perceived that a very grateful odour came from these trees, which spread all over the island...
>
> I felt great pleasure in admiring the island, which is a very pleasant place. Here it seems that nature has spared nothing to render this isle delightful above all others that I have ever seen. It is very well disposed for the support of man, having wood and stone and lime for building houses, and wanting only labourers to cultivate these fine plains where one finds salt in abundance, while the coast swarms with fish. There one hears the chatter of the birds, which make these odorous woods resound with their sweet songs.[1]

The party returned with fresh herbs and samples of the island's many plants. At sunset they noted 'a great number of fires burning the whole length of the coast on the mainland'. If their arrival had stirred interest among Aborigines on the mainland, there was little to show for it; when parties went ashore, they found only the embers of abandoned camps.

As the Dutch headed for Batavia, they fired their cannon in a final salute to the island that Vlamingh had dubbed 'a terrestrial paradise'. And he promised

These sentinel Rottnest Tea-trees (Melaleuca lanceolata) *and other native trees once covered the island, prompting early visitors to comment on its well-wooded nature. But by the 1920s, when opened to the public, Rottnest had already lost half its tree cover.* Photo by Pat Barblett.

the expedition's maps and charts would provide 'ample proof' of its intensive examination of the Southland.

The legacy of maps the cartographer left behind would be well used by Dutch navigators—and by the British when they established the Swan River Colony. A year later, Vlamingh left Batavia for the return journey to The Netherlands via the Cape of Good Hope.

Vlamingh gave the island its name (originally written as Rottenest) and it was more than a century before his 'wood rats' were correctly identified as marsupials by a naturalist aboard a French vessel that visited in 1801.

These early years of exploration not only gave the island an identity, but fostered one of its lingering legends related to an unmarked tombstone that can still be seen in the Rottnest cemetery. If the tomb had an inscription, it has long been eroded by the elements—and that was perhaps sufficient to ensure the survival of the story told by early residents.

The weathered tomb is said to mark the grave of a ship's officer—some say French, others Dutch—who was buried by a remorseful young rival following a fatal duel. The survivor later returned to the island to build a tomb for his compatriot. The French version has a capricious woman at the centre of a feud that simmered and then flared during a long and difficult voyage to Australia.

Historical archives have revealed nothing to substantiate the hearsay; however, islands have always spun legends—and perhaps this particular one owes as much to its setting as the romantic notion of a man who died for love far from home.

4. BIRTH OF THE SWAN RIVER COLONY

FRENCH interest in Australia in the early nineteenth century clearly acted as a spur to imperial Britain, then riding a tide of exploration, conquest and trade that would secure it the largest empire history has known.

In 1822 the brig HMS *Bathurst*, under the command of Captain Phillip Parker King, anchored north-east of Rottnest. The captain had been despatched to report on the region's prospects, and while doing so his botanist, Allan Cunningham, noted the cypress pines and tea-trees on Rottnest, and the absence of many of the characteristic trees of the mainland.

Five years later another British expedition examined the prospects of establishing a colony in the Swan River area of Australia's west coast. Captain James Stirling noted Rottnest on his charts, but his chief focus was on the mainland area he would encourage his government to claim. Stirling believed that great opportunities awaited settlers risking their lot in another outpost of Britain's expanding empire.

The first free settlers of the Swan River Colony arrived on the barque *Parmelia*, which rounded Rottnest Island on 2 June 1829. The enthusiastic and resourceful Captain Stirling was the obvious choice as governor.

For settlers, the appeal of the Swan River Colony lay in the generous land grants offered, and the spirit of buoyant optimism that accompanied the venture. Stirling himself became a land holder as well as administrator, being granted some 4,000 acres between the Swan and Helena rivers, where he developed his estate, Woodbridge. In the confusion that developed over the complicated business of land allocation, it was not long before settlers appealed to the Surveyor General's office to issue grants on Rottnest Island.

Being uninhabited added to Rottnest's attractions. Relations between land-hungry settlers and the Aborigines were deteriorating, for the newcomers were quick to punish those hostile to their conquest.

There was eager talk about how the island could be developed, and the townsite of Kingstown was carefully planned in 1831. More than a hundred small lots and several larger holdings were initially offered, and one of the first settlers to take up land was Robert Thomson.

Thomson arrived in the colony in 1829 and took possession of 1,000 acres on the Swan River and, later, several lots on Rottnest. He planned to farm his island estate and collect salt from the salt lakes.

Thomson had high hopes of building a secure home for his wife and family. He had feared for their safety on the mainland during the two years he spent in Fremantle, so after a spell of running the Stirling Arms and a ferry service on the Swan, he moved his wife and eight children, their furniture and piano to Rottnest. Perhaps he dreamed of making his fortune, of duplicating the fine estate in Scotland that his wife, Lady Caroline Lydiard, had given up when she had eloped with the romantic master of her father's hounds.

While Rottnest seemed to offer unlimited opportunities for farming and fishing, it soon became clear that life would be a constant struggle for the Thomsons. Land was cleared, limestone was used for building, crops were planted, but the much-hoped for prosperity eluded the large family. Soon Caroline Thomson was writing to the Governor and Colonial Secretary asking for help in selling the family's grant of land on the Swan—and appealing for bags of wheat.

'Fremantle, with Rottnest Island in the distance, 1859', watercolour copy by W. S. Hatton from engraving by E. Y. W. Henderson, Rex Nan Kivell Collection. The National Library of Australia, nla.pic-an5924537.

> Fishing is so very uncertain as it is now five weeks owing to the weather we have not been able to go out consequently there is no provision for the family...I can assure you the life we live here is most miserable with respect to food...At the present we have nothing but greens to eat...Under all circumstances I hope you will pardon my taking up so much of your time...[2]

The announcement in 1838 that the island was to become a prison for Aboriginal offenders, proved the final straw for the dispirited Thomson. By this time his health and sight were failing. He was offered, in compensation, a land grant at Cookernup, where he died in 1865, aged 73. Thomson's dreams of an island estate had foundered, but perhaps by the time he died his bitter memories associated with the island had faded.

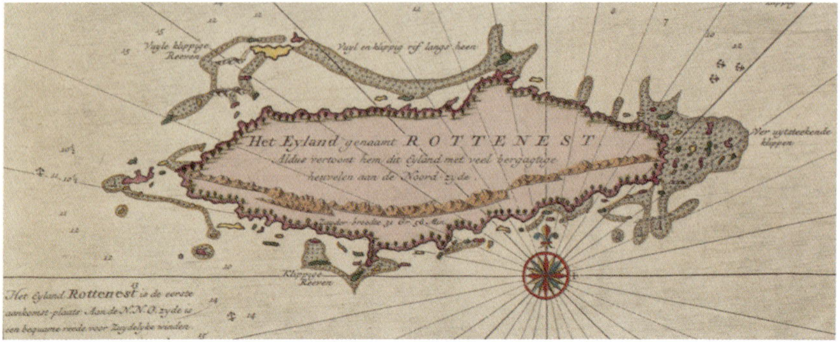

Map of Rottnest by Johannes van Keulen, 1753. The National Library of Australia, nla.map-nk2111.

5. THE ISLAND PRISON

THE Nyungar people would have known Rottnest as a long limb of land projecting from the coast before it became an island some 7,000 years ago. The rising ocean radically altered Western Australia's coastline, submerging landscape features that marked hunting grounds or fishing rocks once favoured by the Nyungar.

The people wove stories about the time when the chain of three scattered islands had been part of the mainland. It was told that a great fire leapt through the thickly wooded country at the base of three ridges, and so fierce were its flames that the earth split and the sea rushed in, isolating Rottnest, Carnac and Garden islands.

During the nineteenth century, the Nyungar would have noted the increase in sailing ships moving from horizon to shore. When, in 1829, Captain James Stirling, the Swan River Colony's first governor, issued a proclamation offering to protect the indigenous people from 'fraudulent, cruel or felonious' actions, they must have sensed the dawning of a new era. Their lives would never be the same.

As settlers struggled to survive and develop the generous land grants they had voyaged half-way across the world to claim, the Nyungar were daily dispossessed of traditional hunting grounds and waterholes. Conflict became inevitable.

Bushland was cleared, fenced and stocked with sheep and pigs, or was ploughed to cultivate vegetables, vines and wheat. Access to waterholes that had sustained generations of Aborigines was suddenly denied. Kangaroos, swans and other local animals were hunted with guns, and their numbers immediately began to dwindle.

The Nyungar had no conception of private property or the accumulation of material wealth when they speared the sheep or cattle that now grazed their hunting grounds. Quickly their numbers multiplied in the dark recesses of prisons that were among the fledgling colony's first public buildings. It was the start of a long and bitter struggle that would see the loss of land, culture and dignity. Fine young men fell in unequal contests that pitted spear against musket. Others were arrested and confined to prison cells in chains, their loss keenly felt by communities suddenly without hunters, fathers and sons.

Entire communities were also afflicted by a string of hitherto unknown illnesses that the settlers brought—influenza, measles, smallpox and venereal disease. They took an enormous toll on the tribal people of Britain's newest colony. As menfolk died or disappeared behind the forbidding walls of prisons, communities became vulnerable. The close-woven fabric of a society shaped and sustained by the land began to unravel.

To the men who suddenly found themselves in chains—for stealing a sheep, being intoxicated by the white man's liquor, or settling a tribal dispute in the time-honoured fashion with a spear—the closing prison doors were an unimagined horror. Death sometimes seemed preferable to the lingering despair of imprisonment for men whose lives had centred on family, community and their spiritual connection to the land. Some leapt from the boats that carried them to prison, others simply withered and died in the cells.

Those who fastened the chains and locked cell doors noted the deep despair that incarceration invoked. The more humane suggested an open prison where inmates could at least work without chains, learn to cultivate crops, build houses and roads, and hunt for recreation.

Official eyes turned to the offshore islands—to Carnac where, for a while, the warrior Yagan was held, and to Rottnest, where Robert Thomson was struggling to harvest crops of wheat and barley. In 1838 the first prisoners arrived on the island—and within days they were escaping to freedom in Thomson's whaleboat. Governor John Hutt quickly appreciated that the island could not accommodate prisoners and settlers, so he resumed the land grants, offering settlers land of equal value on the mainland. In 1841 a statute to establish an open prison on the island was passed.

Governor Hutt's plan for the Aboriginal people in the colony was to 'civilise' and 'assimilate'. Schools for Aboriginal children were established, settlers were urged to train indigenous people so that they could find jobs in the fledgling colony. Those who fell foul of the law and were despatched to Rottnest would 'be instructed in useful knowledge, and gradually trained in the habits of civilized life...'[3]

It was accepted that continued close confinement was prejudicial to the health of Aboriginal prisoners. It was acknowledged that '...the Island of Rottnest appears peculiarly suitable to their detention, inasmuch as a greater degree of personal liberty may be allowed...'[4]

Thomson Bay, named after Rottnest's first settler, served variously as the off-loading point for prisoners, an anchorage for official vessels, and then later, as pictured, the main entry for holiday-makers.
Battye Library 5663B/19.

6. HENRY VINCENT, THE ONE-EYED GAOLER

Three iron bars in a norrow window was all the view Rottnest prisoners had of the outside world. Photo by Pat Barblett.

EARLY reports from the prison's first superintendent, Henry Vincent, painted a positive picture of the penal settlement that was growing daily, for the island's first gaoler worked relentlessly and demanded much of those he supervised. In July 1840 he noted that the prisoners were in good health and he praised their hard work. Three acres of wheat and two of barley had been planted, and the building of the barracks for prisoners was well advanced.

Vincent looms large over the island's early years as a prison. He would serve two decade-long terms as superintendent. From today's vantage point, it is easy to condemn him as a cruel overseer, all too ready to inflict the lash and chain on those who obstructed his authority or attempted to flee from custody. However, his actions reflect the harsh reality of crime and punishment in the nineteenth century, when conditions in gaols round the world were wretched, and harsh convictions were handed down, often for minor offences.

Arriving in the Swan River Colony in 1830, Vincent was appointed gaoler at Fremantle Gaol, but in 1839 he was sent to Rottnest to establish the penal settlement. The superintendent himself knew the harshness of army discipline. He had served with the British Army and, at 18, was wounded, losing an eye in the Battle of Waterloo. When he arrived in the colony, he had a wife and seven children, and a reputation as a severe disciplinarian. He was 42 when he assumed the post of superintendent, and he clearly impressed the colonial authorities.

Vincent not only had the experience and authority to ensure an efficiently run institution, he also possessed a restless energy. He galvanised his prison labourers to plant and plough, so that the institution would, as far as possible, be self-sufficient. He set about building a penal settlement that eventually included an octagonal prison, a boys' reformatory, a chapel, a lighthouse, a salt works, a mill, a military barracks and stables, and accommodation for the island-based pilots and their whaleboat—plus an impressive summer retreat for the colony's governor.

Initially, his own living conditions were basic and the establishment he ran was plagued by shortages—of skilled labour to help him build the settlement, of flour for the prisoners and his own family, of medicine and blankets. He complained in writing to the Colonial Secretary that no cargo boats with supplies or instructions had arrived, that he had received no candles for himself or his troops for two months, and that not a light was to be found on the island. When night fell, he lamented, he was obliged to go to bed or sit in the dark.

When supervising the construction of the first lighthouse in 1842, the civil engineer Henry Trigg provided an insight into the daily life of the prisoners. Cells were unlocked at 5.30 a.m. and the men were immediately despatched to perform tasks. Some cut grass for the animals in the prison farm, others checked the fishing lines, cut wood, quarried stone, or collected fertiliser for the gardens. Those who remained in the prison warmed the breakfast. After the meal, work resumed on the foundations of the lighthouse. Some of the prisoners attended to domestic duties within the prison, while others were sent fishing. After a meagre evening meal the prisoners were permitted to hold a short corroboree.

Trigg recalled the trauma of those who keenly felt the distance that separated them from their communities. He wrote:

> The prisoners will sit down and weep most bitterly particularly old men, or those who have left wives and children on the main; and when they see the smoke from the fires at the place where they have been accustomed to meet when unshackled and free...they seem intensively alive to their lost freedom and lamentably bewail their captivity.[5]

Some of the prison's inmates were legendary figures. Captain, convicted of murder and sentenced to life imprisonment, first arrived on the island in 1891. He was later paroled to a police station in Derby where he teamed up with the elusive Jandamarra. The two were known for their marksmanship, and, when acting as armed assistants to a police officer, they turned their weapons on

him. At the time they were accompanying a party of prisoners being taken to Derby.

Captain and Jandamarra released their prisoners, stole guns and ammunition, and took to the rugged terrain of the Napier Ranges. Jandamarra, also known as Pigeon, was shot dead by pursuing police after evading capture for weeks. Captain was eventually caught, convicted and sent back to Rottnest in 1896. A year later he was dead, one of the many victims of the influenza that swept through the overcrowded cells.

Influenza, measles and other epidemics claimed many lives. When the octagonal prison containing thirty-three cells was completed, sometimes as many as ten prisoners were crammed into a cell measuring 5 by 3 metres. They slept on the floor with blankets.

During both his spells as superintendent (1839–49 and 1856–66), there were complaints about the harsh discipline of Henry Vincent. While inquiries cleared the superintendent, his son William, an assistant warder, was convicted of the death of a prisoner in 1859. He spent his three months' term of 'hard labour' working in police force stables, and was later accepted into the force.

In 1866 Vincent retired, after thirty-six years in the colonial government. He was old and infirm, and at that late stage the colonial government saw fit to express displeasure at a brand of discipline it had tolerated for two decades. They reduced his pension and Vincent died in obscurity.

The brutality of the early wars that had shaped him, and the quality of mercy of a colonial regime that tolerated him, made Vincent's term of office a dark era for Rottnest. However, there is one positive legacy: the survival

During Rottnest's term as a prison island, thousands of Aborigines from across the State were sent there—among them, this group in the 1890s. Many never left, as influenza and measles epidemics took a heavy toll on the overcrowded prison population. Battye Library 3045B/224.

The small Rottnest chapel, one of many heritage buildings to arise under Henry Vincent, reflects the Spanish architecture that the former soldier must have admired during the Peninsula War. It closed in 1903, but was rededicated in 1965. Photo by Pat Barblett.

of the many settlement buildings that he designed and that the prisoners constructed.

The simplicity of these limestone buildings, their sturdy construction and sometimes graceful contours, have withstood winter storms and summer heat. They form the historic heart of Rottnest and are admirable examples of basic colonial architecture. Architects involved in its restoration consider this intact ochre-washed settlement rising above its protective seawall to be the only one of its kind in Australia. (The ochre was originally created by adding rusted nails to the lime wash, when whitewash created an unacceptable glare.)

The residence Vincent built for the governor would, when Rottnest became a holiday island, be converted into flats, and in the 1950s, into a hotel. The hay store and grain crushing mill became a museum; the schoolhouse and chapel remains a place of worship. The boys' reformatory now provides holiday accommodation, as does the octagonal prison.

During Rottnest's term as a prison island (which officially ended in 1904), some 3,700 men and youths from across the State were incarcerated in the gaol that witnessed five executions. There is an Aboriginal cemetery on the island and seventeen Aboriginal heritage sites have been listed. Perhaps the future awaits an appropriate island monument that recognises the hardship of those who looked towards the mainland from *Wadjemup* and wept.

7. AN EMINENT VISITOR

HENRY Vincent was succeeded by several rather more benign superintendents, the first of which was William Jackson, who had served as the Rottnest pilot since 1857. Jackson became superintendent in 1867 and invested a great deal of energy into making the island productive. Fruit and vegetables flourished around the salt lakes while wheat, barley and rye were tended on the slopes of Mount Herschell.

The superintendent ran sheep and pigs, bred horses for sale to the police force and sold salt to the mainland. He also ushered in an era of less rigorous discipline, though he failed to address pressing problems such as overcrowding and unsanitary conditions.

In 1872 Rottnest received a visit from the celebrated English novelist Anthony Trollope, who recorded that there were sixty-five Aboriginal prisoners who received 12 ounces of meat a day 'properly cooked with rice and bread and tea'. On Sundays they were 'allowed to roam at will through the island and they bring home wallabies and birds and fishes'.[6] Snakes were plentiful on the island and prisoners were offered a reward of tobacco for every snake killed.

During his visit Trollope witnessed what he considered to be the best corroboree he had seen in Australia. However, he noted that even this high-spirited event was tainted by the situation of those who danced in the prison courtyard.

> When the order was given I could not but think me of other captives who were desired to sing and make merry in their captivity. Here was no unwillingness and when I proposed that 5s. worth of tobacco should be divided among the performers I was assured that the evening would be remembered as a great occasion.[7]

It took an outbreak of influenza in the 1880s to draw attention to the prison's cramped, unhygienic conditions. Eighty died, more than a third of the prison's population. Seven weeks later the Governor, Sir Frederick Napier Broome—who enjoyed visiting Rottnest to relax and shoot black duck and teal—ordered an inquiry. By that time nearly all the surviving occupants had contracted measles.

The death toll shocked the mainland community. The *Inquirer* newspaper called the treatment of prisoners unnecessarily severe and claimed that the prison was not being properly run.

The Surveyor General, John Forrest, chaired the commission whose prominent members, including the Colonial Surgeon, Dr Alfred Waylen, all expressed shock at the unsanitary conditions. They advocated more blankets and clothes be given to prisoners. The inquiry concluded:

> ...we are unable to recommend its [the prison's] abolition...it has disadvantages, among which we may mention the cold in winter, and the want of fertility of the soil...The great advantage of this Island Prison is that the natives can be worked without chains and that they are allowed liberty on Sundays to roam...[8]

In the year of the inquiry, Jackson's successor was announced—William Henry Timperley, a former police inspector. He would serve from 1883 to 1890. By the time he assumed office, the island also accommodated a boys reformatory which housed young settler offenders aged from 8 to 17. The inmates included two 13-year-olds who were sentenced to three and six months respectively for stealing cakes and pigeons.

The superintendent's son, L. C. Timperley, left vivid descriptions of the arrival on the weekly mail boat of apprehensive, chained prisoners 'with long hair and beards matted together with lime, charcoal and grease',[9] and of happier occasions such as the weekly corroborees.

> Those taking part painted various designs on their bodies with lime, manufactured spider web head dresses from rushes, sticks and the woollen bindings from their blankets…On these occasions small fires were lit in the courtyard...The rhythmical stamping of the natives' feet in the bolder forms of corroboree could be heard all over the settlement.[10]

He also wrote of the occasional escapes by new arrivals, who were soon tracked down: 'With so many expert black trackers on the Island, the runaway's liberty was always short lived.'[11]

8. SENTINELS OF LIGHT

FOR those who knew Rottnest as a prison, the ocean that surrounded the island was a source of misery, daunting and isolating. However, for the men of empire charting the development of a colonial outpost, the sea lanes that linked Fremantle with the trading world were an economic lifeline.

Early Dutch navigators were the first to identify Australia's west coast as treacherous, and they made particular mention of the dangers of landing at Rottnest 'on account of the reefs of rock along the coast'.

The island was soon dubbed 'the brick on the doorstep of Fremantle'—ready to trip unwary sea captains tempted to celebrate the sight of land with a cheering measure of port. By the turn of the century, the island had already claimed twelve vessels.

The Wadjemup lighthouse, one of the island's most enduring landmarks, still casts its light across island and ocean. Officially opened in 1896, it significantly improved the safety of merchant shipping approaching Fremantle. Photo by Pat Barblett.

Decades on, more accurate charts and greatly improved navigation aids were still no insurance against disaster. With the prize almost in her grasp, the sleek French ketch *Anitra*, front-runner in a 1979 international race from Plymouth to Fremantle commemorating the establishment of the Swan River Colony, ran aground off Rottnest. The irony was palpable; 150 years earlier the barque *Parmelia*, bearing Captain James Stirling and the first settlers, had narrowly escaped the same fate when she grounded in a storm off Garden Island.

In 1840 the importance of establishing safe shipping lanes to Fremantle was acknowledged by the decision to build Western Australia's first stone lighthouse on Rottnest. Limestone was quarried and convicts laboured to erect the 20-metre tower. With its welcoming beam (powered by coconut oil and later kerosene) reaching across 18 nautical miles, it provided a vital link between land and sea.

Towards the end of the century a second lighthouse, twice as high, was called for, and the more powerful light from its lanternhouse became something of a proud symbol for the remote 'Cinderella' colony.

The 1880 and 1890s were decades of unprecedented development. The discovery of the world's richest square mile of gold-bearing earth transformed the Swan River Colony and greatly increased the colony's shipping. Within a decade the population of Western Australia doubled and an economy once dependent on wool and timber, sandalwood and pearls, suddenly had a resource guaranteed to spur development and prosperity.

Railways, a water pipeline to the Goldfields, a modern port to welcome merchant shipping to Fremantle—all were products of the gold boom that increased trade tenfold. However, vessels approaching Fremantle did so with caution, knowing the hazards. In the two decades prior to the turn of the century, some 1,000 ships had come to grief off the coast, so the Premier, Sir John Forrest, believed that a new lighthouse for Rottnest was vital.

Sir John was determined that the harbour being built by his brilliant Chief Engineer, C. Y. O'Connor, would lure the large mail and passenger steamers away from Albany's natural harbour, till then their main port of call in Western Australia. Fremantle would be the hub of maritime commerce and instructing O'Connor to build a bigger and better lighthouse at Rottnest meant that the colony now possessed a trinity of welcoming beacons to rival the world's best—at Albany's King George Sound, Cape Leeuwin at Augusta, and at Rottnest.

In the autumn of 1896 a vice-regal party crossed the waters to Rottnest for the official opening of the lighthouse that—along with Fremantle harbour, railways and the Goldfields water pipeline—would be O'Connor's enduring legacy. After travelling to the island in a paddle-steamer, the officials were subjected to a 'rough and bumpy' ride in drays driven by prisoners.

As a fresh sea-breeze whipped the colourful flags decking the lighthouse's impressive 37-metre tower, it was proclaimed that the 45,000-candlepower light was 'not only a light to guide the mariner to our shores, but also a magnificent milestone on the road to the progress of the colony'.[12] It was further suggested that the lighthouse was a symbol that would 'lead the people of the country on to great and noble deeds and thoughts'.[13]

Such rhetoric was echoed by many over an excellent luncheon, as the cream of colonial society warmed to the task of welcoming the fact that a 'new light had dawned on Western Australia'.[14]

OPENING THE NEW LIGHTHOUSE AT ROTTNEST ISLAND W.A

The official opening of Rottnest's main lighthouse in 1896 hailed it as a symbol of progress for the colony. Battye Library 20967P.

9. FOR THOSE IN PERIL ON THE SEA

WATCH the advance of an armada of purple clouds beyond the white tower of Bathurst lighthouse, and the surging surf as a south-westerly whips the island, and it is easy to imagine the horror of being caught in an angry sea in the wooden barques that sailed to and from Western Australia in the nineteenth century.

The peril was not always confined to the stricken vessel; rescuers also risked their lives, often in dramatic circumstances. When the barque *Lancier* struck a reef south of Rottnest during a gale in 1839, the island-based pilot, Captain James McLean Dempster, rushed to its aid in a whaleboat. As the *Lancier* began to break up in high seas, Dempster, with the help of Aboriginal prisoners, brought passengers and crew safely ashore.

Vain attempts were made to save the most precious piece of cargo—an iron chest of coins valued at £7,000—but it slipped between outstretched hands as the vessel was buffeted, and was lost to the ocean. In recognition of the gallant rescue, the ship's owners presented the pilot with a fine silver bowl with an inscription that noted 'the noble and liberal hospitality' offered after the vessel's ill-fated arrival from Mauritius.

Sandalwood was one of the colony's first exports. In China it was turned into furniture, and sweet-smelling incense sticks that were burnt at temple shrines. With a cargo of sandalwood bound for Shanghai, the barque *Lady Elizabeth* left Fremantle in 1878, sailing swiftly into a storm. The barque tried to turn for home, but rolling seas swept it onto a reef. Defying howling winds and the black night, Rottnest's pilot lugger struggled to reach the wrecked vessel, but was forced to turn back. Attempting again as dawn lightened the dark sky, the pilot succeeded in rescuing all of the crew. Once the storm had abated, the crew managed to salvage most of the precious sandalwood swept ashore, but so ferocious was the winter ocean that there was soon no trace of the vessel.

The competitive nature of shipping also raised the stakes, as happened in 1886 with the 499-ton Liverpool-built *Mira Flores*, chartered by local merchants to challenge British dominance in mercantile trade. The barque carried not only a valuable cargo of heavy machinery and iron for railways, arms and ammunition, beer and spirits, but the hopes of West Australian traders anxious to establish an independent shipping service. A reef cut short this grand plan when the ship was totally wrecked. The WA Shipping Association eventually reached agreement with London broking firms to work together.

Despite its powerful beam spanning 23 nautical miles, the new Rottnest lighthouse did not halt the loss of ships or eliminate that most delicate of factors, human error. In the winter of 1899, a tragic chain of events led to the death of twelve members of the merchantman, the *City of York*, when it struck disaster off Rottnest's north coast. The vessel had been approaching landfall when a flare, confusingly sent up from the island, was mistaken by the captain for that of a pilot vessel. The mistake proved costly; the ill-fated boat struck a reef and one of its lifeboats capsized in high seas. The mishap prompted a decision to build a secondary lighthouse at Bathurst Point, on the north-east corner of the island.

Lighthouses played a vital role in the pilot boat operations by providing a communication link between the pilot boat station and incoming ships. The Rottnest Island Pilot Station operated between 1848 and 1903, when a signal station was established on the island. Once a vessel was sighted, the news was

BOURNE

then telephoned to the lighthouse in Fremantle and a steam-powered pilot boat was despatched (the pilots had previously used double-ended whaleboats). The last boathouse built for pilots in 1859 still exists on the island.

Some thirteen wrecks can be found in the waters surrounding Rottnest. Some can be explored with snorkel, others with diving equipment. The Rottnest Wrecks Heritage Trail brochure lists those that can be visited and safe practices that must be observed. The trail is a combined venture involving the Western Australian Museum, the Rottnest Island Authority and other sponsors.

Today, as pleasure craft converge on the island, this heritage trail of Rottnest wrecks conjures up a distant and very different era of shipping.

No beacon of hope could save the stricken merchant vessel City of York, *which foundered in stormy seas off the island's north coast in 1899 with the loss of twelve lives.*
'The Wreck of the City of York', by George N. Bourne © Western Australian Maritime Museum.

10. A VICE-REGAL RETREAT

WHEN, in the last decade of the nineteenth century, the number of prisoners declined—by 1900 there were only twenty-six—there was much debate about the financial wisdom of maintaining the prison island. By that time, the State's fortunes had turned with the discovery of gold, and suddenly there was money and time for recreation.

The fact that Rottnest was cooler than the mainland at the height of summer had not escaped the notice of colonial governors. In 1840 Vincent built a comfortable family residence (cottages F and G, main settlement) with French casement windows; however, the governor of the day was quick to acquire it as a summer residence. In the 1850s Governor Kennedy drew attention to the

Government House, built in a commanding position overlooking Thomson Bay, was converted into flats to accommodate visitors when the island was opened to the public. Battye Library 816B/B3639.

deplorable state of this island residence and ordered the building of a 'marine summer residence'. In 1863 the *Perth Gazette* reported that a 'very tasteful and comfortable' Government House was taking shape.

While Kennedy hoped to sample the comforts of the grand residence that Vincent was building on the shores of Thomson Bay, it was not ready for occupation until 1864, by which time his successor had assumed office.

Governor Hampton (1862–68), along with successive governors, enjoyed entertaining members of colonial high society in his summer place. Lady Alice Lovat wrote in 1872: 'I have been staying lately at Rottnest, my country or rather Island home...We were up every morning and out with our guns by 6am and had some good sports, a mixed bag of quails, pigeons and sandpiper. Later in the day we used to go out sea fishing. In fact we had a very jolly week.'[15]

L. C. Timperley, whose father was superintendent, recalled the visits of the Governor, Sir Frederick Napier Broome:

> It was customary for him and Lady Broome with their guests and staffs of officials and servants to spend about six weeks on the Island each summer. His Excellency was a keen sportsman and spent his leisure hours fishing and shooting. A small jetty ran out in front of Government House on which was erected a bathing box for the use of the Government House party.
>
> When the ladies bathed the men were conspicuous by their absence, and vice versa. The Foreshore in front of Government House was out of bounds to the Islanders whilst the vice-regal party was on the Island...

As all game on the Island was protected for the sport of Governor Broome and the vice-regal party, many good bags of the long red legged Rottnest snipe were shot down by those sportsmen, as were many brace of mountain and black duck, also teal. When a 'shoot' was arranged, a party of natives was detailed to various parts of the lakes to drive the snipe towards the sportsmen, who were sheltering behind brushwood brakes at points of vantage round and behind the various lakes. Wallaby drives were also arranged for the vice-regal party, who proceeded beyond the main light-house to Narrow Neck, where the scrub would be set alight by a party of natives who headed as many wallabies as possible towards the guns of the shooting party.[16]

When Responsible Government was proclaimed in Western Australia in 1890, the privileges of governors began to be questioned. The gold boom spawned money for leisure, and extended the number of well-appointed Rottnest cottages for the favoured few, but travel to the island still required a permit.

In 1894 *The West Australian* lamented the fact that access to Rottnest was restricted to a dwindling number of Aboriginal prisoners on the one hand, and a privileged group of high society on the other. The newspaper article followed a proposal by Elias Solomon, a parliamentarian, that no more prisoners be sent to the island and that it should become 'a place of summer resort and recreation'.

Parliamentarians were soon debating the future of Rottnest. One confided that he could conceive of nothing more attractive than owning a summer place on the island. Clearly he was nurturing dreams of a sybaritic retreat

For some early holiday visitors to Rottnest, camping was a popular alternative means of accommodation. Battye Library 816B/B3640.

complete 'with steam boats and pleasure boats going backwards and forwards from the mainland'.

While Premier John Forrest was reluctant to remove the Aboriginal prisoners from the island, the government conceded that it would investigate turning the island into a 'health and pleasure resort'.

11. STRAW HATS AND BEACH PYJAMAS

IN 1902 the decision was made to close the penal settlement, and transform the island into a pleasure resort. However, while the prison officially closed in 1904, prisoners continued to live on the island, and to build roads and repair buildings. In fact, it continued to benefit from prison labour until 1931.

With tuberculosis and other diseases still taking a heavy toll on Western Australia's population, Rottnest was viewed by government officials as 'a sanitorium second to none in Australia'. As early as 1902 local trades unions had given voice to the popular sentiment that the island should be 'dedicated to the people as a public park forever', and that no permits to reside on the island should be extended to anyone other than those whose business it was to 'cater for the public comfort and convenience'.

Holiday-makers arriving in the 1920s disembarked from one of the ferries that plied between Fremantle and Rottnest, there to board a horse-drawn 'tram' or the 1915 T Model Ford run by the hostel. Battye Library 4697B/3 and *(opposite)* Battye Library 5663B/2.

Dr William Somerville, an articulate leader of the local labour movement, championed the cause of 'the people's island' when attempts were made to privatise land. The University of Western Australia Archives.

Governor F. D. G. Bedford backed the notion of a public park and recreation ground and advocated that the natural beauty of the island should remain undisturbed. However, there were those who wanted their own slice of the island that Vlamingh had pronounced 'delightful above all others'. Early governors had given the island their seal of approval, and since 1840 access had been strictly controlled.

The favoured few with access to Rottnest were reluctant to relinquish their retreat to those beyond their charmed circle. Colonial Secretary J. D. Connolly hastily proposed the subdivision of 300 two-hectare blocks for private sale or lease. However, Dr William Somerville, supported by unions and local councils, became the 'voice of the people'.

Somerville had trained as a blacksmith and worked as a boilermaker before arriving in Fremantle in 1895. While becoming an articulate leader of the local labour movement, he also played a significant role in the establishment of The University of Western Australia, serving on the Senate from its first meeting in 1912 until his death in 1954. He was Pro Chancellor of the University from 1936 to 1941, and his creation of the University's atmospheric Somerville Auditorium—now a favourite venue for the UWA Perth International Arts Festival film screenings—reflected an enthusiasm for greening the landscape that would also benefit Rottnest.

In 1907 Somerville put pen to paper in defence of Rottnest. He wrote to *The West Australian*:

> ...in a democratic community it will not be seriously contended that the man who has only means sufficient to erect a humble camp in which to

> spend a summer holiday has as much right to a block if he desires it, as the plutocrat who would erect a mansion.[17]

The proposal to privatise the island would, he said, establish 'a little oasis where our local silvertails can retire, secure in the knowledge that they are safe from intrusion by the common herd'.[18]

In the same year it was proposed that, to preserve vice-regal privacy, the boundary wall of Government House be extended seawards to Thomson Bay and to the shore of Government House lake. When this was refused, the governor of the day, clearly feeling rebuffed, decided to make Albany his summer retreat. (Government House was later turned into accommodation for holiday-makers, and in 1953 became the Rottnest Hotel, known to all as 'The Quokka Arms'.)

In 1917 Somerville's impassioned plea on behalf of the people—and the groundswell of pressure it unleashed—achieved its aim: Rottnest Island was declared an A-Class Reserve. This assured that no portion of the island could be leased or sold except by legislation, and regulations were introduced to try to protect flora and fauna. There was to be no shooting; no horses and no dogs were permitted on the island; and it was made clear that misconduct and 'riotous or indecent' behaviour would not be tolerated.

A hostel (in what had been the reformatory) and an annexe (formerly the old gaol) began to welcome visitors, while a general store and tearoom near the jetty catered for their needs. Government House, the bungalows and cottages all benefited from repairs and a fresh coat of paint.

During the 1920s camping sites were set up in Thomson Bay and Bathurst

Point. Those wanting more privacy trekked along a dirt track to Salmon Bay opposite Green Island, where there were small timber and brush huts, and a well.

As the 1930s dawned, the island's profile as a holiday island seemed secure. Gay processions of weekenders moved between the tiny settlement and the natural pool of The Basin. In line with the relaxed atmosphere, dress codes were jettisoned, and newspapers noted that 'skirts were the exception'. In the summer of 1934 the *Daily News* observed:

> To wear an ordinary street or sports dress at Rottnest this summer, is to live in the past, and to call back memories of when bathing costumes were always covered by wraps, some of them so striking that a wag labelled the path to the swimming basin, 'Kimono Parade'.[19]

A degree of dress code formality prevailed around the time this picture was taken, but by the 1930s skirts became more the exception than rule, giving way to 'beach pyjamas', shorts and slacks. Battye Library 4697B/5.

Fetching straw hats—little coolie hats or wide-brimmed Mexican numbers—sat on immaculately shingled heads, and were worn with beach pyjamas, smart shorts or slacks.

The influx of sunseekers stamped the island as everyone's favourite 'getaway'—but those with dreams of owning a slice of the island continued to work behind the scenes. One parliamentarian suggested that a residence on Rottnest would admit him to 'the charmed circle of refined aristocrats'. In 1932 Parliament prepared to debate a bill to grant leases for building sites on the island for periods up to twenty-one years. There had been only one dissenting voice to this move on the Rottnest Board of Control. It was the island's champion, Dr William Somerville.

The introduction of the bill took the matter out of the boardroom into the public domain. In a letter to *The West Australian*, Dr Somerville pointed out that twenty-five years previously he had canvassed municipal councils, roads boards and other public bodies. They had been unanimous in their opinion that 'this unique and beautiful island—the whole of it—should be preserved as a public playground for the people for ever, and that no man, however influential or wealthy, should be allowed a right to exclude others from any portion of it...'[20]

The proposal to grant leases was withdrawn, and Western Australians gave a collective sigh of relief. Rottnest seemed once again assured of remaining 'a public playground for the people'. Later that year, the board announced the island's best year since it had assumed control. The board received no financial assistance from Government, all of its revenue being derived from

In case holidaymakers needed further persuasion, the Government Tourist Bureau was happy to oblige. This illustration, from a 1927 tourism booklet, captures not only the latest in costumes, but a carefree atmosphere as well. Battye Library 919.412 ROT.

rents, royalties on the salt gathered and bagged from the lakes, landing fees and trading concerns.

By 1935 the hostel was accommodating 200 people, and bungalows, cottages, flats and camps were booked up weeks in advance. Over the Easter weekend, some 1,200 gathered on the island to play tennis, ride, swim, walk, lounge on the lawns of the hostel or beaches, or enjoy the dances at the tearooms.

The ferries *Zephyr* and *Emerald* were kept busy, as were the light planes that had been carrying passengers to Rottnest since the opening of a landing strip in 1930. The natural earth surface of the latter had no lighting facilities for emergency night flights, and on occasions kerosene-soaked rags set in containers lit the small strip.

There had been much excitement when, in November 1930, the first aircraft landed, just a year after air services to the island were pioneered by the Baker Aviation Company using a light seaplane. In 1946 WA Airlines would commence services, in which year 2,679 visitors flew to the island. Within two decades the number had risen to more than 14,000.

However, while visitor numbers were spiralling in the 1930s, the dark clouds of war were gathering in Europe. Rottnest's location as a forward defence position would soon eclipse its value as a holiday island. Soon defence force personnel were looking at the island as a wartime asset, and were making preparations for war, as they had in 1914 when Rottnest became, briefly, an internment camp for prisoners-of-war and European migrants unlucky enough to have come to Australia from an enemy nation.

Excited crowds celebrate an aviation milestone—the late 1930 christening of Rottnest's newly opened airstrip. Such flights provided a speedier alternative to the busy ferries plying between mainland and island.
Battye Library 5663B/3.

12. ROTTNEST'S ROLE IN FORWARD DEFENCE

ROTTNEST'S location made it ideal as a lookout for military and civilian shipping approaching Gage Roads and Fremantle.

As early as the 1880s, when the race for colonial possessions was at its height, it was feared that war between the imperial powers, Britain and Russia, was imminent. So real was the sense of foreboding that a plan was put in place on the island to signal the arrival of the Russian fleet by igniting simultaneous bonfires at the main lighthouse and at Bathurst Point Lookout. Fortunately, there was never a need to warrant such action.

However, when World War I was declared on 4 August 1914, Rottnest's military fate seemed inevitable. Ten days later the Defence Department commandeered the island, and shortly afterwards it was accommodating its first wartime prisoners.

With its role as a penal settlement still looming large in the minds of officials, the island was deemed a suitable location for an internment camp for some 1,700 German and Austrian prisoners-of-war and internees. Among the first were the officers and crew of the German ship SS *Neumunster*. The crew was accommodated in the camping area, the officers in cottages. Armed soldiers on horseback patrolled the island, but the internees were allowed freedom of movement during the day.

In his book *The Home Fires*, Anthony Splivalo writes a moving account of how Western Australian migrants from Europe, considered 'enemy aliens', experienced the war years. Initially, they were interned on Rottnest before being sent to a camp in New South Wales.

Splivalo, who had left his Dalmatian homeland four years earlier to join a brother working in Kalgoorlie, recalls a sunny morning in 1915 when, along with others—Austro-Hungarians, unnaturalised Germans, Turks and Bulgarians (many working on the mines)—he was taken to Rottnest.

Having sorely missed the sparkling sea that washed his homeland, Splivalo found himself occupying one of the tents set up for holiday-makers among the island's pine trees. While young, uniformed military clerks struggled to register unfamiliar names such as Tomljashevich and Jakovljevich, the young migrant lamented that these once-friendly Australians had suddenly been distanced by a conflict half-way across the world.

However, the ache of being 'imprisoned' was diluted by the island that was to him 'extraordinary in loveliness'.

'I shall never forget my first walk over the causeway between two small lakes', he wrote. 'The scene filled me with joy and made me forget the war and the horrible bitterness it had stirred in people's hearts.'[21]

The men were allowed to roam the island by day, discovering its coves and fishing spots. The Germans romped naked, sunning themselves with an abandon that shocked the more conservative Slavs. With several excellent cooks in the community, the daily catch became a welcome addition to rations, and the aroma of bouillabaisse or barbecued fish settled over the camps in the evening. Rations could also be supplemented by items from the general store.

Splivalo received letters from the Australian Kalgoorlie family with whom he had lived. He wrote letters for many of his companions who, unlike him, were less fluent in the language. His language skills also won him the task of reading out the news of the day to those who gathered in his tent.

However, late in 1915 the camp was disbanded and the prisoners were transferred to one in New South Wales. Nearly 1,500 men regretfully packed their belongings and boarded a transport vessel, rightly anticipating that their new camp would not offer them the freedoms of Rottnest—a fear confirmed by the sight of its barbed wire perimeter fence patrolled day and night by soldiers.

Decades later, when World War II loomed, one of the guns from the original HMAS *Sydney*, which had sunk the raiding German cruiser *Emden* off the Cocos Islands in November 1914, became a vital part of the Rottnest defences.

Gunners of the 6th Heavy Battery, Royal Australian Artillery, with one of the two 6 inch guns at Bickley, c. 1938. Photo reproduced courtesy of the Rottnest Island Authority (88-172).

13. WORLD WAR II

WELL before the declaration of World War II, the Army began building Kingstown Barracks at the eastern end of Rottnest. A narrow-gauge railway ferried building materials from the jetty to the building site where the barracks, administration block, hospital, officers' mess, canteen, cottages for NCOs, workshops and store began to take shape.

When war was declared on 3 September 1939, it was clear that the island's location made it a valuable forward defence facility and that the preparations under way had already transformed the holiday island into Rottnest Island Fortress. Sweeping ocean views from Wadjemup Hill had become a vital asset that would earn the island the title of 'gatekeeper'. Guns deployed on it would be able to engage approaching enemy vessels before they could attack Fremantle Fortress, the second line of defence.

When, in 1940, the island was closed to the public and declared a prohibited place under national security regulations, *The West Australian* speculated that while summer holiday-makers would be disappointed, the restrictions could well be lifted 'before the beginning of the next tourist season...'

At that stage no one expected a long war, or that within a year Perth City Council would be constructing air raid shelters as fears of invasion mounted. After Japan attacked Pearl Harbour in December 1941, its triumphant troops swiftly moved down to claim Singapore. Fighter planes then appeared in the blue skies over Broome and Wyndham in March 1942, convincing many that it was only a matter of time before enemy aircraft and naval vessels would be threatening the capital city and its port.

On Rottnest, camping areas that had once accommodated attractive young girls in beach pyjamas and Mexican hats now saw their former escorts bivouacking in areas around Bickley Point and Oliver Hill.

The Bickley Battery was armed with two 6 inch guns of naval origin that had been converted for land use. The guns had a proud provenance, coming from the original HMAS *Sydney*, and from HMAS *Brisbane*. When these vessels were scrapped, their guns were salvaged and installed at Bickley Point in 1938. They fired 45 kg shells, had a range of 24 kilometres and could fire up to twenty rounds a minute. When it was discovered that a local landmark, Phillip Rock, obstructed the Bickley searchlight and the easterly field of fire, the rock was quickly cut down in size.

At Oliver Hill the battery boasted two 9.2 inch guns with a range of 30 kilometres and a 360 degree arc of fire. They were capable of engaging a battleship, and during training, their firing of high explosive shells reverberated through the island. A warren of underground connecting tunnels and facilities at Oliver Hill accommodated plotting rooms and magazines.

Together the heavily camouflaged batteries were capable of targeting unwelcome enemy vessels approaching Gage Roads and the harbour entrance through the South Passage. A brick four-storey fortress post on Wadjemup Hill—still an impressive landmark—had the best vantage point and directed all the gunfire from the fixed artillery positions in the Northern Fire Command which comprised batteries on Rottnest and the mainland.

Soldiers from 'civvy street' stationed on Rottnest declared its living conditions to be good, if you wanted a 'silent war'. Initially, boredom made many impatient to see action; however, Pearl Harbour changed all that. In 1942 some 2,500 personnel were stationed on Rottnest. The Royal Australian

Navy manned the wooden signal station (now refurbished) that played a role in controlling military and civilian shipping out of wartime Fremantle. The Royal Australian Airforce provided an early warning and meteorological unit.

When calls to action saw many stationed at Bickley depart for overseas, the Australian Women's Army Services personnel replaced departing soldiers at the instruments in both batteries and the Fire Command, with thirty personnel stationed on the island during 1942.

The batteries and railway had been camouflaged, and mock guns and a dummy railway were constructed. When the USS *Pinegrove* arrived in Fremantle in 1942 with an anti-aircraft regiment, guns were deployed to protect the battery position from strafing aircraft and dive bombers.

Fortunately, the guns of Rottnest were never fired in anger and have now

The World War II battery at Oliver Hill, its underground tunnels restored, is one of only three locations in the world with a 9.2 inch gun installation still intact. Photo by Pat Barblett.

'Home' may not have been that far away, but wartime mail was still important to these Australian Women's Army Services recruits stationed on the island between February 1943 and November 1945.
Photo reproduced courtesy of the Rottnest Island Authority (02-078).

become a tourist attraction. Once the war was over, Rottnest's champion, Dr William Somerville (who had frowned upon the military presence on the island, particularly the careless destruction of trees), began pressuring for it to be reopened. However, having had a taste of Rottnest, the Army was reluctant to totally relinquish its holdings.

Somerville rightly anticipated that 'there are too many with a vested interest in retaining the present position for them to easily relax their hold of the Island'. In April 1945 troops were withdrawn and the Army ended its control of the island, with the exception of an area around Bickley.

In July 1945 all restrictions on visiting Rottnest were lifted, but it took several months of work by 200 Italian prisoners-of-war before a clean-up was completed. The Italians removed barbed wire, filled in trenches, replanted trees and cleaned up the camps and cottages. (During the war Italian internees, rounded up a day after Italy entered the war, had been briefly accommodated on Rottnest before being transported to a camp at Harvey and then to South Australia.)

In 1946 the batteries were dismantled and the two guns at Bickley were sold to a scrap merchant. Just over forty years later, the gun barrels, which had been buried on site, were returned to the island as the battery at Oliver Hill, with its underground tunnels restored, became another Rottnest attraction.

Railway enthusiasts gave their time to restoring the wartime railway which followed the original 7 km track. Visitors can now board a train, the Captain Hussey (named after the engineer who managed the installation of the guns) and take a ride to Oliver Hill before exploring the underground combat tunnels. Rottnest is one of only three locations worldwide with a 9.2 inch gun installation still intact (the others being Gibraltar and Robben Island, South Africa).

The island's signal station, which played a crucial role in Fremantle's coastal defences, has also been refurbished. The station, an integral part of the island's extensive military heritage, was opened in 1937 on Signal Ridge and used to control military and civilian shipping out of wartime Fremantle.

In 1984 Kingstown Barracks was handed back to the State Government, and a year later became the Environmental Education Centre seeking to heighten public awareness of the island's unique historical and natural heritage. Today Rottnest Island Education Services offer diverse programs for schools, teachers, community groups and members of the public. Groups are accommodated in the old barracks and many participate in community service activities related to the island's conservation.

Hatted against the sun, artillerymen at Bickley clean a 6 inch gun. Photo reproduced courtesy of the Rottnest Island Authority (89-054).

14. LAND AND WATER

SOME 18,000 years ago Western Australia's coastline lay 12 kilometres west of Rottnest and the sea surged around it some 130 metres below today's level. However, melting glacial ice caused the ocean to rise.

Geologists believe that more than 7,000 years ago Rottnest was still part of a long limb of land stretching out from the mainland near Fremantle. Rising waters reshaped it as an island 6,500 years ago, then further submerged it, creating a suite of ten islets. At that time, the areas now occupied by the salt lakes were arms of ocean reaching between the islets. When, some 5,000 years ago, the sea level fell, the contours of the island we know today emerged. Initially, the shallow pools of the lakes remained linked to the ocean, but eventually the rise of dunes and beach ridges isolated them.

The rise and fall of the ocean has left a trail of 200 islands off the coast of Western Australia. Some became protected breeding grounds for birds. The quokka and the venomous dugite snake survived on Rottnest and Bald islands, while the tammar wallaby grazed the vegetation of Garden Island.

Rottnest encompasses some 1,900 hectares, is about 11 kilometres long and 4.5 kilometres at its widest point. Its highest point is Wadjemup Hill which rises 50 metres above sea level, and its 40 km coastline comprises rocky limestone headlands and fine white sandy bays protected by limestone reefs.

Rottnest's limestone-sculpted bays are its greatest asset. Parakeet Bay, with its sweep of fine sand, is a favourite beach for swimmers—and those who wish to capture the essence of the island in paintings. Watercolour by Jan Grainger.

The rugged, layered limestone that defines the island's bays evolved as sand dunes, rich in lime from myriad shell fragments and tiny marine organisms. Time, and the action of rainwater, cemented this material into limestone. Water channelling through it carved subterranean caverns, and those in the area of the salt lakes collapsed. When the rising ocean separated Rottnest from the

The salt lakes of Rottnest are fringed by unique vegetation, and are a rich source of food for birds. Photo by Pat Barblett.

mainland, it flooded the collapsed cave systems to form the tranquil lakes now occupying a tenth of the island.

As migrating birds fly over Rottnest's low-lying heath and undulating groves of trees, the sheen of the island's hypersaline lakes instantly signals to them the presence of food such as insects and crustacea. The salt lakes are also attractive to visitors, for on a windless day they perfectly reflect the sky and shoreline, adding another dimension to the landscape.

There are seven permanent lakes (Government House, Serpentine, Herschel, Garden, Baghdad, Vincent and Timperley) and five smaller lakes (Pink, Negri, Sirius and the two Pearse lakes). Their water is four times saltier than the sea and a salt industry was one of the island's first, back in the 1830s. Robert Thomson collected salt from the Pearse lakes and sold it on the mainland, and salt continued to be collected until 1950. The island's salt works and store were demolished in 1959, but their foundations can still be seen at the Causeway's western end.

The level of water in the salt lakes fluctuates by more than a metre seasonally. Summer shrinkage exposes sandbars and a gently sloping shoreline that is a rich feeding ground for birds. Winter storms replenish them to mean sea level, merging some of the smaller lakes to the fringe of salt-tolerant vegetation that girds them.

Blue-green algae mat the base of the lakes, while algae containing beta-carotene give the smaller lakes their pinkish hue, and also colour the red brine shrimps that feed on them. (Elsewhere in Western Australia, beta-carotene is harvested from salt lakes and exported to Japan as a food colouring and vitamin supplement.)

Hardy, salt-tolerant plants growing on the fringes of the lakes have evolved their own survival strategies: the samphires concentrate salt in red 'beads' that shrivel and die, while saltbush expels salt through its leaves. Both were harvested as vegetables by early settlers.

Lack of fresh water has been a perennial problem on the island. Freshwater seepages around the lakes discharge underground water from aquifers, and quokkas and birds are dependent on these for survival.

After the island became a reserve attracting visitors, roof catchments on the hostel, shops and cottages were supplemented by two roofed underground storage tanks, and wells with water of varying degrees of salinity. A rainwater catchment area, originally sealed with bitumen by the Army in 1939, was expanded in the 1960s to provide water for the settlement, in addition to that brought from Fremantle by barge.

When new settlements at Longreach and Geordie Bay were planned in the 1970s, a serious hunt for water began. Early artesian bores and wells had produced nothing but brackish water. However, a re-evaluation of groundwater prospects by government geologists located significant volumes of potable supplies, including a 10 metre thick lens of water in the limestone west of Wadjemup Hill. This find meant that in 1976 the water barges ceased running.

However, a reliable supply of potable water remains one of the most important management issues facing the island, and visitors are urged to conserve water. In recent years two desalination plants have helped the island meet its needs without adversely impacting on its freshwater aquifer and the environment.

15. PLANTS OF ROTTNEST

Swathes of dainty blue lace flowers, or Rottnest Daisies, carpet the grassy plains in spring, adding their distinctive colour to the palette of the island, and attracting the attention of pretty jewel beetles. Watercolour by Elizabeth Rippey.

WHEN rising oceans isolated Rottnest, the combined effects of exposure to salt-laden winds and limited fresh water took a toll on vegetation. Seven thousand years ago, the island was a eucalypt woodland. However, majestic stands of tuart, marri, jarrah, banksia and sheoak disappeared, and scientists estimate that perhaps 90 per cent of plants were lost. Now only fossil pollen collected by scientists remind us of that very different landscape.

However, hundreds of hardy native species did survive the transition to an island environment, for Western Australia's rich flora has evolved countless strategies for surviving both bushfires and dry summers. Though the region receives more rainfall than London, most of it falls over three winter months.

On Rottnest, survivors such as sea rocket, pigface and spinifex colonise beach areas, and wild rosemary scents the air. Sheltered from the ocean, the low scrub is patterned with creamy clematis in spring, golden acacias, and the red of templetonia. On open plains, tussock grass is interspersed with mauve and yellow stilt plants, and swathes of blue lace flowers (Rottnest Daisies).

Native trees that predominate include Rottnest Tea-trees (*Melaleuca lanceolata),* the small dark Rottnest Island Pines (*Callitris preissii*), graceful Native Willows (*Pittosporum)* and gold-flowering *Acacia rostellifera*.

Soon after the island became an A-Class Reserve, tuarts, peppermint trees, oleanders and palms were introduced. These, and other exotic species, have become well-loved landmarks: the avenue of Moreton Bay fig trees in the settlement offers a tunnel of deep shade in summer, while Norfolk Island Pines form a familiar backdrop to settlement buildings.

The first botanist to visit the island, Allan Cunningham, noted in 1822 that the Rottnest Island Pines were abundant throughout, 'occasionally relieved by…melaleuca…and the more elegant pittosporum'.[22] But with settlement, the forest was cleared for building and farming. By the 1920s, when camping sites were being established, the demands of the prison era had become evident. Trees cut for fuel, and fire used as a hunting tool, had robbed the island of half of its tree cover.

Since settlement, several devastating fires have swept through areas of the island, reducing the soil's seed bank of pines and tea-trees, both being sensitive to fire. Acacias that regenerate after fire gradually replaced pines and tea-trees, but they in turn were restricted by the appetite of the island's only marsupial, the quokka.

When Rottnest became a reserve, its resident wallaby was declared a protected species and, in following decades, it multiplied dramatically. The quokka's favourite food is *Acacia rostellifera*, which had dominated the island while the marsupials were hunted, but was heavily grazed once they became protected.

When early attempts were made at reafforestation, this herbivore confirmed that it was game to graze almost any plant except the Prickle Lily. Fencing saplings became the only answer: eucalypts and tea-trees required protection for three years and pines for up to ten.

The island's management has, for decades, attempted to make good the vegetation losses of the past. In the 1930s Dr William Somerville, a member of the Rottnest Island Board, was an enthusiastic pioneer of greening, propagating

The distinctive Rottnest Tea-trees (Melaleuca lanceolata) *are shaped by the strong south-westerlies that fan the island.* Watercolour by Jan Grainger.

trees in an island nursery. Some 2,000 plants, including all the island's natives, plus Port Jackson figs, date palms, South Australian wattle and South-West peppermints, were nurtured.

When Somerville retired, his efforts were advanced by other committed individuals. In an oral history interview Rottnest Island Board member Thomas Sten described how he once nurtured as many as 500 trees at a time, later to be transported to the island on the *Wandoo* freight boat.[23] He claimed personally to have planted 400 acres of trees over twenty years, and his dedicated work is acknowledged in Sten Park, one of the many areas he had revegetated. The greening efforts of local schools and others are also recognised in successfully rehabilitated areas.

The Rottnest Island Authority continues efforts to revegetate the island, but today only plants native to the island are used. No attempt is made to green the entire island, but rather to focus on low-lying areas at the eastern end that are fenced and planted, or are merely allowed to regenerate naturally.

16. THE QUOKKA (*Setonix brachyurus*)

QUOKKAS were initially mistaken for large rats, thus prompting Dutch explorers to give the island its name.

Once widespread on Western Australia's mainland, quokkas were hunted, faced food competition from rabbits, and were preyed upon by introduced foxes, cats and dogs. By the 1960s there were few left on the mainland, so populations on Rottnest and Bald islands are important.

Among the first marsupials recorded by early explorers, the quokka is one of the smallest wallabies. Whereas its preferred habitat is a swampy thicket, this herbivore has adapted to Rottnest's limited water supply and more open environment. The island population fluctuates, particularly groups living in the waterless west, where there is little tree cover.

While some quokkas remain active during the day, most retreat to the deep shade of tea-tree groves, emerging from their hidden lairs at twilight.

When freshwater soaks are inaccessible, quokkas eat water-bearing succulents such as pigface, in addition to grasses and wattle. Because the plants they graze are deficient in nitrogen and other important trace elements, quokkas can become undernourished and anaemic, causing some joeys to die towards the end of summer and in autumn.

When conditions prove too harsh for a joey nestling in its mother's pouch, a stand-by embryo develops, one that would otherwise wither. The young are in the pouch for four to six months, and remain dependent on mother's milk for several more. The Rottnest female then waits till the following spring to breed again, unlike mainland quokkas, possibly because feeding her offspring has depleted her reserves.

Quokka numbers vary, but are thought to stand at about 10,000. The marsupials have had an enormous impact on the island's vegetation, and their increase in numbers has been partly responsible for vegetation decline. There is a continuing debate among scientists as to whether the population can be managed to allow the plants on which they survive to regenerate (see The laboratory island, pp80-1). Scientists suggest that the quokka's poor quality diet may predispose it to infection by salmonella, which is widespread in the island population. The quokka appears to be particularly vulnerable to infection when under environmental stress during summer.

Scientists have concluded that the proliferation of salmonella in the Rottnest quokka represents 'a significant reservoir of infection' because of the degraded habitat. They suggest this situation could prevail as long as 'seasonal starvation' remains the dominant feature of the biology of the Rottnest quokka.[24]

Rottnest's emblematic quokka (Setonix brachyurus), *usually emerges from its hidden lair at twilight. Once hunted, but now protected, the population has multiplied to more than 10,000, taking a toll on the island's vegetation.* Photo by Pat Barblett.

17. ROTTNEST—ON THE WING

ROTTNEST'S diverse habitats—exposed reef platforms and offshore limestone stacks, salt lakes and freshwater swamps, coastal heath and woodlands—make it quite unlike any other island off the Western Australian coast, and a protected haven for birds.

Some are residents—like the busy Welcome Swallow that nests in the eaves of your cottage or the bold Singing Honeyeater that hops onto a window ledge looking for snacks. Others include seasonal migrants from as far afield as Siberia, introduced species, or those that simply arrived and stayed.

The Peafowl and the Ring-necked Pheasant were introduced by early governors for ornament and sport. The Laughing Turtledove established itself on Rottnest in the 1930s, while the colourful Rainbow Bee-eater arrived in the 1970s and has bred on the island ever since. The island's melaleuca woodlands are now home to the Red-capped Robins and Golden Whistlers, both of which are no longer present on the Swan Coastal Plain.

Some 112 species have been recorded over or around the island, 49 species occurring regularly, including 14 transequatorial migrants that breed in the northern hemisphere, then fly south to escape winter.

While their breeding grounds are snowbound, these transients spend several months feeding on the island's salt lakes. Rottnest is just one of several protected bird habitats in Western Australia that welcome around two million international visitors annually. Some—like the Red-necked Sandpiper—weigh no more than 30 grams, but annually make this epic 26,000 km round trip.

The birds fly from Siberia, northern China and Mongolia via South-East Asia, arriving in Australia in August and September. Thousands descend annually on

Facing page: Banded Stilts (or Rottnest Snipe) rise from a salt lake. Photo by Pat Barblett. *Above: The colourful Common Pheasant (or Ring-necked Pheasant) was initially brought to the island in the 1920s.* Watercolour by Elizabeth Rippey.

Western Australia's tropical north, with its milky turquoise water and emerald mangroves. The huge inter-tidal flats of Roebuck Bay and Eighty Mile Beach are part of a region that offers the highest biodiversity of marine organisms in the world. It is an ideal place to restore depleted reserves, and Roebuck Bay has been declared a Wetland of International Importance under the Ramsar Treaty, which recognises and protects the habitats of migratory birds.

While some of the migrants spend the northern winter in the State's tropical north, others continue their journey south to Rottnest and other favoured habitats.

Pied Cormorants nest in colonies and are a familiar sight, perched on rocky outcrops or drying outstretched wings in the sun. The birds feed on fish and crustaceans and are able to speedily pursue their prey underwater.
Photo by Pat Barblett.

They arrive on the island between September and November desperately in need of a good feed. Some select small crabs and marine worms, others burrow for crustaceans and bivalves, eating them shell and all. Non-breeding migrants, including the Bar-tailed Godwit, Sanderling, Ruddy Turnstone and the small Red-necked Sandpiper, forage in flocks of several hundred along the margins of the salt lakes.

When they head north in March and April, they need to have doubled their body weight to be able to fly non-stop for three or four days. They leave their summer feeding grounds when the northern spring thaw begins and food supplies are once again available.

Long-distance visitors include the Whimbrel, its down-curving beak prising out a veritable feast—molluscs, crustaceans, worms, beetles, flies, moths and spiders—after its Russian diet of berries. The Curlew Sandpiper, another migrant that breeds in north-eastern Siberia, also uses its black bill to good effect as it builds up its body weight for the long return journey. This wader is easy to spot during high summer as it forages on Government House and Baghdad lakes. It returns to its Siberian breeding grounds in April.

One of the most characteristic bird calls on Rottnest is the high-pitched call of the Osprey, or Fish Hawk, that wheels high above ocean and coastal heath, before falling to seize a fish or a lizard.

In Western Australia, Osprey hunt mainly in coastal areas, nesting on rocky headland or limestone stacks. There are several pairs resident on the island. A restless bird of prey, the Osprey is forever hunting for food for its mate and chicks. Its tools of trade are a sharp hooked beak, strong curved talons, piercing

The Osprey (or Fish Hawk) is a striking sight, hovering stationary over potential prey in the ocean or coastal heath. The birds build distinctive nests of sticks and matted seagrass on rocky headlands and limestone stacks.
Photo by Pat Barblett.

eyes and wings strong enough to hold it stationary before dropping feet first to snatch its prey.

Prior to eggs being laid in August, the birds are busy gathering sticks to repair their prominent nest of twigs and debris gathered from beach and heath. Ospreys return to their nests year after year, adding new material. They mate for life and display to one another, renewing their bonds, before producing eggs.

The two or three chicks born in October remain in the nest for just under two months, and towards the end of this period they approach the nest's edge, encouraged by the calls of adult birds wheeling above. Soon the entire family is exchanging calls across a blue bay while hunting above the reefs—but not for long.

Juveniles fly off to secure their own coastal territory, and young banded birds have been recovered up to 500 kilometres away within months of leaving Rottnest. Other global habitats have been far less kind to these magnificent birds of prey; their empty nests have become grim reminders of their decimated numbers, and the devastating effects of pesticides.

Equally spectacular when hunting is another resident, the Nankeen Kestrel. This small falcon also hovers stationary over its prey—a lizard or skink, a small bird or baby dugite—its rufous wings outstretched, its white breast catching the sun. Then it drops to its quarry before flying off to a springtime nest in the crevice of a cliff face or an offshore stack. There are several breeding pairs on the island.

The kestrel is a familiar sight close to Little Parakeet Bay, and as you watch it, you invariably also hear the distinctive cry of the Common Pheasant in the

Once a rare visitor, the Red-necked Avocet now resides on the island in small numbers. Using its upturned bill to search salt lakes for brine shrimp, the chestnut-headed bird is fiercely protective when hungry gulls wheel above its downy chicks.
Watercolour by Elizabeth Rippey.

An industrious pair of territorial Pied Oystercatchers forage along the shoreline. These resident birds adeptly use their long, probing bills to dislodge molluscs and marine worms.
Photo by Pat Barblett.

heath around Baghdad lake. A handsome bird introduced by former governors looking for sport in the 1920s, the pheasant is a native of Asia Minor. With its blue head, red-masked eyes, russet body and long tail, the male is eye-catching as it takes to the air when disturbed while hunting for snails and ants in the heath. It has a distinctive call that makes it easy to locate.

In sandy coves you inevitably encounter the Pied Oystercatcher, using its probing red bill to dislodge molluscs and marine worms. This bird often shares a sculptured sweep of white sand with a colony of Crested Terns that spends much of the morning preening in the sunshine. However, as the day advances, these attractive birds with crested crowns, spotless white necks and dashing yellow bills go on fishing sorties across the bay. Elegant in flight, the terns will keep you entertained with aerial displays of formation flying, spectacular solo dives and elaborate courtship dances.

Come springtime, the island's ornamental Mountain Ducks (Australian Shelducks) draw attention. These residents of the salt lakes and swamps are handsome in their green, white and chestnut plumage. They mate for life and in spring proud parents may be seen leading a platoon of ducklings from nesting sites in limestone crevices or offshore islets to their all-important brood territory.

Brood sites need to have a supply of fresh water and be close to the lakes and swamps where the ducklings will feed. Only a few dozen sites exist on the island, and they are fiercely defended. Leading newly hatched two-day-old chicks from a nesting site to this territory is an epic journey fraught with danger. The sky is patrolled by hungry ravens, and even when the chicks reach the sanctuary of the lakes, Silver Gulls will try to isolate vulnerable offpsring from

Mountain Ducks (or Australian Shelducks) establish a fiercely guarded brood territory in swamps and salt lakes before nesting on headlands and cliffs. When the ducklings have hatched, they are led in procession to this territory—either swimming or walking—while Silver Gulls and ravens hover above.
Photo by Pat Barblett.

the brood. Even the most watchful of parents lose 20 per cent of their brood before the chicks learn to fly away from trouble.

Loss of chicks is an even more serious problem for the Banded Stilt, or Rottnest Snipe, the bird that once provided sport for the governor's shooting parties. These dainty black and white birds with long pink legs are a familiar sight, feeding on the lakes. They congregate in tightly packed rafts, and take to the air in unison when disturbed.

Initially, it was thought that the stilts went abroad to breed, for their eggs were never discovered on the island. However, we now know that they breed on distant mainland salt lakes, and the brine shrimp on which they feed only hatch when these lakes are flooded by rare storms. Scientists don't know how the stilts detect these storms in the dry interior, but when they occur, thousands abandon Rottnest within a day.

The stilts fly inland in flocks and nest on islands and spits in these salt lakes. Although their chicks quickly fatten on brine shrimp, increasing numbers of marauding gulls can decimate both eggs and chicks. On Lake Eyre the slaughter was such that in recent years National Park rangers have culled the gulls to ensure the preservation of the stilt population in south-eastern Australia. Fortunately, as yet Western Australia's inland lakes appear to be free from the gull problem.

The West End of Rottnest is a favourite destination for visiting birdwatchers. With torches ready, they await the spectacle of undulating dark waves of Wedge-tailed Shearwaters returning to their burrows after the sun has set. There are several growing colonies of the birds, which arrive on the island in August to breed; from then till April the night air is filled with extraordinary orchestrations of their nocturnal song. They are true seafarers, leaving the island before dawn to spend the entire day feeding, gliding above the waves and sometimes diving ten metres in pursuit of prey. The breeding migrants are truly industrious when building or repairing burrows that will keep their chicks safe while they are at sea. A pair may shift some 82 kilograms of soil to create a 4-metre burrow that accommodates a single egg, laid in November and hatched in January.

The nestling leaves the burrow in April to join the adult birds at sea. By May the network of burrows at the West End has fallen silent, as the shearwaters fly north, following the sun. Researchers estimate that more than 11,000 nest on the island, their numbers having doubled in a decade. They also speculate that this tropical seabird—mainly seen in the State's mid- to north-west—appears to be extending its breeding areas and established colonies further south, due perhaps to global warming.[25]

With so many migratory birds visiting the island, it is no surprise that the Rottnest Island Authority is considering the benefits of registering the island's wetlands with the Ramsar Convention, to which Australia is a signatory.

Rottnest is one of many habitats Australia-wide that is part of a Birds Australia Wader Study Group. Members of Birds Australia visit the same beach each February to report on the number and species of waders recorded over a weekend. Recently, during an annual visit, members of the group discovered several families of the Painted Button-quail, a small chestnut bird that had been recorded on the island many years ago, but had not been seen for some time. Their find was yet another example of a bird that, despite the island's limited fresh water, appears to have found a home on Rottnest.

The Wedge-tailed Shearwater spends the day cruising above the waves, sometimes settling in vast rafts. The seafaring birds congregate at the West End where the female lays a single egg. Watercolour by Elizabeth Rippey.

18. THE OCEAN'S HARVEST

WASHED by the Indian Ocean and contoured by bays and rocky headlands, Rottnest is a peerless playground for Western Australians.

Far from the distractions of television and playstations, children rediscover the natural world—a baby octopus stranded in a rock pool by retreating tides; secretive crabs emerging from rock crevices; or the undulating menace of a stingray that lies motionless in the sand before erupting in a flurry of movement. Close at hand, their parents may be setting cray traps on a promising offshore reef or casting a line that might provide a feast of dhufish—the island's choicest eating fish—to compliment a bottle of Western Australian wine.

Some 350 species of fish have been recorded in Rottnest's waters, ranging from colourful tropical reef-dwellers swept down the coast on the warm Leeuwin Current, to migratory mackerels and marlins that are summer visitors.

Because Rottnest's waters are 4 degrees warmer than those washing the adjacent mainland from May to September (when the Leeuwin predominates), tropical fish survive the cooler winter months, and some ninety tropical species have been recorded. They are particularly drawn to coral reefs in protected areas of water on the southern side of the island.

The waters off Rottnest are reserved exclusively for recreational fishing. Most of the fish caught by anglers are migratory, and include herring, tailor, whiting, skipjack, garfish, marlin and tuna. The Australian herring is probably the island's best-known species, especially in cooler months when northward swimming schools are seen in local waters, sometimes followed by the distinctive dorsal fin of the grey nurse shark.

Crabs are many and varied, and enthusiasts are particularly delighted to net some tasty blue manna. Rock crabs, with olive carapaces and orange claws, are common, as are the pale gold ghost crabs that scurry across the beach at night in search of food. Red hermit crabs swiftly retreat into shells that they carry aloft when feeding; and seaweed crabs camouflage their bodies with seaweed as they forage.

For many visitors who know the best location for traps, the prize is the western rock lobster, found only along Western Australia's continental shelf. A prosperous export industry centres on these slow-growing lobsters, which can achieve impressive proportions—but there are restrictions on their capture by amateurs.

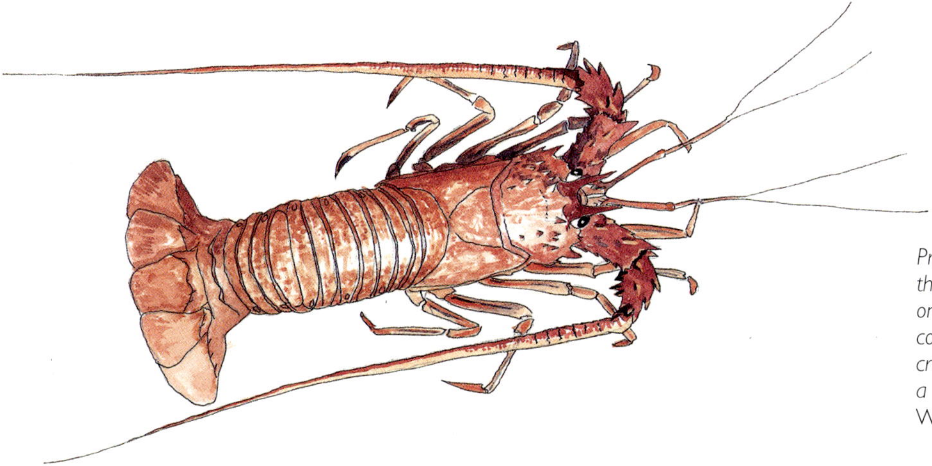

Prized as a culinary delight, the western rock lobster is found only along Western Australia's continental shelf. The slow-growing crustacean is seasonally trapped for a prosperous export industry. Watercolour by Elizabeth Rippey.

Extensive seagrass meadows serve as a nursery for many species, and as a habitat for cobbler, leatherjacket, flathead and the carnivorous seapike drawn to this rich larder.

Rottnest is also a playground for dolphins. The arrival of a school of bottlenose dolphins—one of eight dolphin species in Western Australian waters—stirs as much interest as the sighting of the humpback whales that move past the island on their annual migration. Whale migration patterns have developed over thousands of years, and humpback migrations are among the longest known in the animal kingdom. Before embarking on their journey up the coast from the Antarctic, these restless giants feed for several

months, then survive on stored fat. They don't feed again substantially until they return to Antarctica.

Population data collected by marine scientists suggests that there is a resurgence of the whales once hunted off this coast. Commercial whaling was banned by the International Whaling Commission in 1966, and it is estimated that some 5,000 humpbacks move along this coast each year.

Whale-watching is now a thriving tourist attraction, with the migrating whales sometimes treating watching vessels to a spectacle of spy-hopping (rising out of the water, head first and then sinking) and breaching (leaping clear of the water before crashing down).

Rottnest waters are reserved exclusively for recreational fishing and offer good catches of tailor, sea garfish, whiting, Australian herring and skipjack trevally. Watercolour (opposite) by Elizabeth Rippey, photo by Pat Barblett.

19. THE LABORATORY ISLAND

BECAUSE of its unique environment, scientists have for many decades used Rottnest as a base for research and learning in a range of disciplines including zoology, botany, geology, history and marine science.

Two noted scientists from The University of Western Australia, Dr Ernest Hodgkin, who combined teaching in entomology with graduate research in marine zoology, and Zoology Professor Horace ('Harry') Waring, both turned Rottnest into a laboratory for themselves and their students.

Writing about Professor Waring in *Campus at Crawley*, the late Emeritus Professor Fred Alexander noted the irony of the fact that in the 1950s little research had been done in Australia on the nation's symbol, the kangaroo. Dr Waring's avowed aim was to 'monograph work on the quokka' so that it would become 'the best known animal in the world...'

A research station was established on the island in 1953 and was used by visiting academics and students from local universities and the Fisheries Department. It was accommodated in a building on Wadjemup Hill near the main lighthouse. However, when this building (which had housed female naval officers during World War II) was recently restored as a heritage attraction, the research station moved to an alternative site near the airport. Today, researchers from most of the State's universities use this as a base for projects that will provide a better understanding of the island's unique environment.

Perhaps the most dramatic body of Rottnest-related research over the past half century involved a UWA postgraduate student, Byron Kakulas, who began working with captive quokkas in the 1960s. Before the neuroscience researcher began feeding paralysed quokkas vitamin E tablets, the world assumed that

The Basin, with the Bathurst Lighthouse on the horizon, has always ranked as the island's most popular and accessible swimming spot. Enveloping limestone platforms create sheltered pools of crystal clear blue-green water. Watercolour by Jan Grainger.

once muscle had degenerated, it was incapable of regeneration. The study's astonishing results changed that thinking and became the starting point for far-reaching medical advances in a range of neuromuscular disorders.

Professor Kakulas went on to establish the Neuropathology Department at Royal Perth Hospital, one of the few centres in the world that offers a comprehensive and integrated molecular genetics service. He is a past director of UWA's Centre for Neuromuscular and Neurological Disorders which attracts scientists from Europe and Asia eager to learn the gene therapy techniques pioneered by the Centre. As a result of its research, the incidence of muscular dystrophy has been cut by two-thirds in Western Australia.

UWA Zoology Professor Don Bradshaw is among several researchers who have conducted field studies on Rottnest's quokka, revealing a high level of salmonella infections during the summer months. A former member of the Rottnest Island Authority's Research Committee, he believes that a smaller quokka population should be the aim of those managing the island's environment.

'I have always been in favour of using research as a means of finding solutions to intractable problems. The problem with the Rottnest quokka is that the population is too high, is in poor condition over the dry summer, and is currently regulated by starvation and death of the young and infirm in autumn', says Professor Bradshaw.[26]

'There have been many suggestions about managing the population over the years. Culling has always been considered unacceptable and the only other option is to revegetate the island. We need to return Rottnest's vegetation to something like that of Garden Island. This would mean more trees—but less shrubs on which the quokkas graze. The result over time would be a smaller, but healthier population. We took over 2,000 samples from tammars on Garden Island, where the tree canopy is intact, and never had a single positive isolation for salmonella. Quokkas in poor condition, as they are on Rottnest, are simply more susceptible.'

Because of its vital role in shaping the climate of Western Australia, the Leeuwin Current is also being studied by UWA researchers. Professor of Coastal Oceanography, Chari Pattiaratchi, says that without this warm south-flowing current, Western Australians would not enjoy the 'good life' aspects of their State: 'We would not have the higher winter rains or the climate, the wines or the lobsters. And there would be no coral reefs in Rottnest, and far fewer tropical fish and plants.'[27]

Researchers are studying the Leeuwin's interaction with continental shelf currents and are also gaining a better understanding of the processes at work in the Rottnest canyon, one of the world's largest submarine canyons. Marking the ancient route of the Swan River prior to the rising of sea levels, the canyon begins at a depth of 50 metres and falls to 5,000 metres. It is an area extraordinarily rich in krill, annually attracting schools of pygmy blue whales.

'During summer up to twenty whales may be found at the site, feasting on ten tonnes of krill a day', says Professor Pattiaratchi. 'We want to find out whether there is a correlation between the canyon, the physical oceanography and the biological productivity of the region.'

The fact that Rottnest was attached to the mainland before being isolated by rising seas makes its geology of great interest to researchers and visitors. The island boasts sites of geological significance contained in the Register of the National Estate and is significant as the largest of a chain of limestone islands and reefs on the continental shelf near Perth.

At least once a year, Associate Professor David Haig, of UWA's School of Earth and Geographical Sciences, takes marine geology students for a study weekend. 'Rottnest lies on the outer continental shelf which represents one of the world's largest subtropical marine platforms covered by calcium-carbonate sediment that was formed by the skeletons of organisms', says Professor Haig. 'The island is a perfect site for this study because, coming under the influence of the Leeuwin, it contains more tropical elements than the adjacent mainland.'[28]

The island's marine platforms are also of interest to those studying populations of invertebrates. Dr Jane Prince, of UWA's School of Animal Biology (who is officer-in-charge of the Rottnest Island Research Station), monitors the health of abalone, limpets, cowries and sea urchins and other invertebrates, in collaboration with Associate Professor Bob Black.

'Because our study is long-term, we can see patterns in the populations and relate them to oceanographic or weather events such as El Nino', explains Dr Prince. 'In 2002, for instance, we recorded massive deaths on the platforms. El Nino tends to weaken the strength of the Leeuwin Current, which results in lower sea levels. This, combined with low tides, exposes large areas of rock platforms and can result in big die-off events. With climate experts predicting more frequent El Nino events, there could be implications for Rottnest in relation to the recruitment of invertebrates and fish—because a

A glowing twilight washes over the crescent of villas in the sweep of Geordie Bay.
Photo by Pat Barblett.

strong current is vital to replenish the island's marine populations.'[29]

Whereas a steady stream of research papers has emanated from Rottnest over the past half century, those charged with administering the island have not always made use of the information. Now there is an appreciation that researchers and the Rottnest Island Authority must work together, and that sound scientific research must underpin decisions affecting the island's fragile environment and the pressures that come with its status as Western Australia's favourite holiday island.

The Authority concedes that there are research gaps to fill, such as the need to monitor fish stocks and to gauge the environmental impact of nearly 900 boat moorings on the water quality (which suffers during peak periods) and on seagrass meadows. Perth boasts the highest per capita boat ownership in Australia, and more than 150,000 of the island's annual visitors arrive in private boats.

The RIA is committed to developing Rottnest as a model of sustainability. Its avowed aim is to ensure that the island's resources and experiences are available for future generations. Its management plan emphasises the social value of Rottnest: 'This social value is often called the "Rottnest Island ethos" and can be described in terms of the self-directed, simple, nature-based and family-oriented experience, very much dependent on a high degree of natural amenity and space.'[30]

The Authority also confirms that the island should offer a high level of access for Western Australians, underscoring a widely held view that Rottnest is, first and foremost, for locals.

20. CHEQUES AND BALANCES

THE Rottnest Island Authority faces enormous challenges. It is charged with maintaining the unspoilt character of the island, while accommodating increasing numbers of visitors—currently some 500,000 annually, a quarter of which are day trippers. Visitors generally express a high level of satisfaction with their sojourn on the island.

Those who benefit financially from rising visitor rates—from ferry operators to local businesses—are delighted at the buoyant visitor statistics, while others are concerned that numbers at peak periods put unacceptable pressures on the island. The Authority concedes that when arrivals peak at Christmas and Easter, there can be negative impacts on the island's environment and social amenities, and excessive demands on resources.

Several government departments have particular roles to play on the island. The Department for Planning and Infrastructure is responsible for all boating regulations and for the maintenance of gazetted roads. The Department of Health monitors environmental health standards. The Western Australian Police Service operates a station on the island. The Department of Environment, Water and Catchment Protection controls marine and terrestrial pollution and advises on environmental protection.

Other departments have an interest in the island as part of their overall brief in this State: the Department of Conservation and Land Management in terms of conserving all of the State's flora and fauna; the Department of Fisheries, which manages the State's fish, marine and aquatic resources; the Western Australian Museum, which preserves significant cultural, natural and maritime heritage sites, including shipwrecks; the Department of Indigenous Affairs, which administers Aboriginal sites of significance; the Heritage Council, which provides for the conservation of the State's cultural heritage (many Rottnest sites are listed on the Western Australian Heritage Register); and the Western Australian Tourism Commission, which is responsible for the promotion, development and marketing of tourism.

Given the passion most Western Australians feel for Rottnest, there is no shortage of volunteers willing to invest their time and energy on the island.

Tree-shaded promenades wind through cottages in Thomson Bay, many dating back to the early years of the island's settlement. The distinctive ochre colouring of the cottages was originally created by adding rusted nails to the lime wash.
Watercolour by Jan Grainger.

There is a long tradition of involvement in greening the island that goes back to nurseries of saplings nurtured by Dr William Somerville, and the RIA does much to encourage such endeavours. The Authority has introduced an admirable community service program that involves schools and community groups in activities such as plant propagation and tree planting; monitoring of seasonal shoreline movements, dune erosion and beach litter; and participation in bird counts and seed collection of native species.

Others contribute to preserving the island's built and natural environments by joining one of several active groups. For more than fifteen years, members of the Rottnest Voluntary Guides Association have been sharing their knowledge and love of the island with visitors. Every day they conduct lively tours that range from West End sunset trips to bird walks. Other enthusiasts join the Rottnest Island Foundation, involved in projects such as reafforestation, construction of board walks and fencing. A useful lobby group for the island has been The Rottnest Society, which also focuses on the island's natural environment and provides hands-on support for tree planting, weed eradication and other environmental projects.

In recent years, Rottnest's generation of revenue has not kept pace with the cost increases associated with operating and maintaining the island. The RIA gains revenue from accommodation, business licences and admission fees—the latter forming part of the ticket price charged by ferry operators. The government provides funding for capital works, such as the restoration of cottages, and for environmental emergencies such as storm damage.

21. ROTTNEST—PAST TENSE, FUTURE TENSE

> *I never hear a crow now, anywhere in the world, without instantly seeing, in my mind's eye, the morning sunlight on a white, Rottnest road. I can see the fat mountain-duck bobbing on the embryo ripples of the salt lakes; and the tiny delicate stilt, scarcely more tangible than the salt crust blown from the water. I can feel again the sudden, seductive coolness of one of Vincent's cottages as you come in from the heat...*
>
> *When one is far away from Rottnest, though—perhaps with the cold northern hemisphere rain slicing down on grey roofs and brown pavements—it is those old cottages of Vincent's that come so nostalgically back to mind. In an exile's dream of Rottnest, the sun is always shining. It warms the thick yellow-ochre walls of the settlement...*[31]

Many decades distance these nostalgic words—but the sentiments are as fresh today as ever. When he settled in Western Australia in the 1920s, Oxford-born Bernard Kirwan Ward, a well-respected local journalist and author, came to know Rottnest well. Over the years, he gathered a cherished suite of island memories that he would bring to mind when working far from his adopted homeland.

Like Kirwan Ward, most Western Australians have their own Rottnest memories, and these have contributed to the iconic status the island enjoys today—and the fiercely protective instinct that makes one of the most

common local newspaper headlines in relation to the island: 'Hands off Rottnest'.

This appears whenever the Rottnest Island Authority announces a grim financial forecast for the island, and hard-nosed commentators insist that it must work on strategies to enable it to 'turn a profit'. At such times, developers and tourism operators are quick to enter the fray. Rottnest's inability to balance its budget, they suggest, would become a distant

The island's salt lakes capture the last colours of a sunset as another day winds down and lights begin to glow at Rottnest Lodge.
Photo by Pat Barblett.

memory were it only thrown open to private investment—perhaps an upmarket resort for well-heeled international tourists could be the island's economic salvation?

When local newspapers are filled with such suggestions, one senses the ghost of William Somerville stirring, and nudging the elbow or a phalanx of letter writers who spring to the island's defence. Why? Because most Western Australians share collective childhood and rite-of-passage memories of escaping across the water to the smudge on the far horizon.

Like most islands, Rottnest imparts a heightened sense of freedom—from familiar patterns of life, from traffic, from parental scrutiny. Everyone has a favourite season. For some, it is high summer when bays are crammed with all manner of craft, from sleek yachts to venerable dinghies with many a 'fishy' tale to tell. At such times, the sky is perpetually blue, the ocean aquamarine, the Basin filled with swimmers. The hotel is bustling, every unit on the island is full, there are queues outside the aromatic bakery, and for the young...well, those legendary parties on secluded beaches!

For others, the island is best enjoyed in winter, when summer's throngs are, thankfully, a distant memory. At such times, the island seems to elongate and exhale. When the winter sky darkens and the stormy ocean surges high on beaches, soup simmers on the stove, and there are books and card games in front of a glowing fire. On fine days, there may be a glimpse of a migrating humpback whale, or even an exhilarating dip in the ocean. Spring and autumn also have their particular appeal, offering benign conditions, uncrowded beaches and the ebb and flow of migrating birds.

Anyone who has ever spent time on an island knows the particular allure of being separated from the everyday world by a stretch of blue ocean. For the first couple of days you buy the newspapers to keep abreast of the events that fill distant newspapers and television screens. But, as each day passes, their relevance slips away.

As one world retreats, another comes into focus. You begin to notice things—the elaborate preening ritual of the black and white terns that share your favourite beach and entertain you with displays of diving; the subtle colour changes in the ocean as the sun rides higher in the sky; children playing with a kite through a long, golden afternoon, their laughter an accompaniment to your afternoon siesta; and the particular calls of birds that mark the passing of the day. And you make time to savour the sunset that takes your breath away—whether viewed from a craggy limestone cliff, or in the reflected colours of a salt lake, its shoreline bathed in mellow light.

'Isn't this the way life was meant to be lived?' you ask yourself, charting a whole new regimen of resolutions that may well not survive the return journey to the city. But recognising their value is as important as your latest collection of Rottnest memories that will linger long after you have picked up the threads of city life.

How do we make sure that the special pleasures we derive from this island are preserved? The answer may well lie in William Somerville's philosophy of viewing Rottnest for its intrinsic values, and in resisting

A stiff breeze, gulls cawing and towels drying on the veranda rail are all part of a holiday on Rottnest.
Painting by Nellie Crawford.

pressure to 'market' and 'upgrade' an island that, for Western Australians, is defined by the simplicity of its accommodations and its unspoilt environment.

In the twenty-first century, resisting those pressures—which a former island CEO concedes are ever present—will be every bit as challenging as 'saving' the island from its own success. The RIA is committed to developing Rottnest as a 'model of sustainability', an admirable aim that (with political will) could see the island become a showcase for renewable energy options and best-practice land management. This quest for sustainability will surely also involve addressing the negative impacts of too many people and boats at peak periods.

Rottnest has always been much more than just a holiday destination, and to equate its administration to that of a resort is unrealistic. It may never pay its way because of the ongoing costs of maintaining its intact heritage precinct, repairing the neglect of the past, restoring the woodlands and protecting a fragile environment. Certainly appropriate commercial rents should be demanded for accommodation, but if Rottnest still fails to turn a profit, most Western Australians would consider their 'investment' in this unique island to be well spent.

Australia's largest State, Western Australia, is blessed with many natural treasures—from the filigree of its northern coral reefs to the wilderness character of its sweeping southern coastline. Where such places exist, there will always be the attendant shadow of exploitation and inappropriate development. Western Australians now appreciate that 'people power' can be a potent tool, whether directed at protecting Ningaloo Reef, the State's old growth forests—or saving a unique island from exploitation. The optimism and energy of those who champion the island's cause will surely persist, for the battle is ongoing. By stirring the shadow of William Somerville, we let him rest easy in the knowledge that 'the people's island' will be handed intact to future generations.

SOURCES

UWA *The University of Western Australia*
UWAP *University of Western Australia Press*

NOTES

1 Phillip E. Playford, *Voyage of Discovery to Terra Australis by Willem de Vlamingh 1696–97*, Western Australian Museum, Perth, 1998
2 Prue Joske et al, *Rottnest Island, a Documentary History*, Centre for Migration and Development Studies, UWA, Nedlands, 1995, p25
3 From 'Act No. 21, 4 & 5 Victoria, 1841, constituting Rottnest a Legal Prison', in R. J. Ferguson, *Rottnest Island: History and Architecture*, UWAP, Nedlands, 1986, p14
4 ibid.
5 C. T. Stannage (ed.), *A New History of Western Australia*, UWAP, Nedlands, 1981, p93
6 A. Trollope, *Australia*, P. D. Edwards & J. B. Joyce (eds), University of Queensland Press, Brisbane, 1967 (first pub. 1873)
7 ibid.
8 Ferguson, pp60–1
9 L. C. Timperley, 'Notes on Rottnest Island', quoted in Joske, p126
10 ibid.
11 ibid.
12 L. Layman & C. T. Stannage, *Celebrations in Western Australian History:* Studies in Western Australian History, vol. X, 1989, Centre for Western Australian History, UWA
13 ibid.

[14] ibid.

[15] Lady Alice Lovat, *The Life of Sir Frederick Weld,* GCMG: *A Pioneer of Empire*, London, 1914

[16] Timperley 'Notes', in Joske, p108

[17] Joske, p191, quoting *The West Australian*, 17 May 1907

[18] ibid.

[19] Joske, p243, quoting *Daily News*, 12 Feb. 1934

[20] Joske, p241, quoting *The West Australian*, 5 Dec. 1932

[21] A. Splivalo, *The Home Fires*, Fremantle Arts Centre Press, 1982

[22] L. J. Pen & J. W. Green, 'Botanical exploration and vegetational changes on Rottnest Island', *Journal of the Royal Society of Western Australia,* vol. 66, parts 1 and 2, May 1983, pp20–4

[23] Chris Jeffery interview, Battye Library Oral History 286, cited in Joske, p251

[24] J. B. Iveson & R. P. Hart, 'Salmonella on Rottnest Island: Implications for public health and wildlife management', *Journal of the Royal Society of Western Australia,* vol. 66, parts 1 and 2, May 1983, pp15–20

[25] author interview, Wes Bancroft, *Uniview*, vol. 23, no. 1, Feb. 2004, pp10–17, UWA

[26] ibid.

[27] ibid.

[28] ibid.

[29] ibid.

[30] Rottnest Island Authority

[31] Bernard Kirwan Ward, *Rottnest Island Sketchbook*, Rigby, 1969, p61

FURTHER READING

The author acknowledges the following works consulted during research for this book:

Alexander, F., *Campus at Crawley*, UWAP, Nedlands, 1963

Cairns, L. & Henderson, G., *Unfinished Voyages: Western Australian Shipwrecks 1881–1900*, UWAP, Nedlands, 1989

Green, N. & Moon, S., *Far From Home, Aboriginal Prisoners of Rottnest Island 1838–1931,* UWAP, Nedlands, 1997

Hutchins, B., *A Guide to the Marine Fishes of Rottnest Island,* Creative Research, Perth 1979

Playford, P., *Guidebook to the Geology of Rottnest Island*, Geological Society of Australia (WA Division) and the Geological Survey of Western Australia, 1988

Rippey, E. & Rowland, B., *Plants of the Perth Coast and Islands*, UWAP, Nedlands, 1995

Saunders, D. & de Rebeira, P., *Birds of Rottnest*, privately published by authors, Perth, 1985

Watson, E. J., *Rottnest: Its Tragedy and its Glory,* privately published by Watson family, Perth, 1998

ACKNOWLEDGMENTS

I am indebted to the specialists who have written or conducted research on Rottnest, for their histories and field guides have proved an invaluable resource. I include among these the Rottnest Island Authority, whose suite of brochures is both attractive and informative.

The visual art and photography appearing in this book are testimony to the island's inspirational quality, and I am deeply grateful to photographer Pat Barblett AM, and artists Jan Grainger, Dr Elizabeth Rippey and Nellie Crawford, for making their paintings available.

Pat Barblett's contribution is particularly fitting; she was the first woman appointed to the Rottnest Island Board, serving on it and its successor, the RIA, for sixteen years, the last three as chairperson. She is also deputy chair of the Rottnest Foundation, deputy chair of the Conservation Commission, and a member of the Council of the National Trust. In 2003 she was made a Member of the Order of Australia, received a Centenary Medal and was the recipient of the Sir David Brand Medal in the Western Australian Tourism Awards.

Jan Grainger paints and teaches from her Mount Claremont studio and has held solo exhibitions of watercolours, acrylics and mixed media works of Rottnest and many other scenic locations in Western Australia. Her work is represented in private collections in Australia, Japan, the United Kingdom, the United States, Canada and New Zealand. Website: jangraingerart.com.au

Nellie Crawford works from a studio in Fremantle. She is best known for still life paintings, but relishes the chance to pack brushes and paints and cross the water to Rottnest. Nellie is represented by three galleries in the United Kingdom and Ireland, and her work can be seen at Yallingup Galleries in the South West and Gullotti Galleries in Perth.

Dr Elizabeth Rippey has used her natural history watercolours to increase understanding and preservation of the life of the coast. She has produced *Seashells of Southern Africa* (Kilburn & Rippey, 1982, Macmillan) and *Plants of the Perth Coast and Islands* (Rippey & Rowland, 1995, UWA Press), and numerous pamphlets on Rottnest. Her close association with the island led to a PhD on its vegetation.

My sincere thanks also to the Battye Library, the National Library of Australia, the Rottnest Island Autority, the National Trust and the WA Maritime Museum for use of images from their collections, and to Kathy Tuppurainen, Dr Jenny Bevan, Colin Campbell-Fraser, Ross and Barbara Haig, John Douglass and the staff of UWA Press for their friendly professionalism.